DATE DUE

DEMCO 38-296

THE ECONOMICS OF
MONETARY INTEGRATION

Third Revised Edition

The Economics of Monetary Integration

Third Revised Edition

PAUL DE GRAUWE

Oxford University Press
1997

t Clarendon Street, Oxford OX2 6DP
New York
angkok Bogota Bombay
Cape Town Dar es Salaam
Delhi Florence Hong Kong Istanbul Karachi
Kuala Lumpur Madras Madrid Melbourne
Mexico City Nairobi Paris Singapore
Taipei Tokyo Toronto
and associated companies in
Berlin Ibadan

Oxford is a trade mark of Oxford University Press

Published in the United States by Oxford University Press Inc. New York

© *Paul De Grauwe 1997*

*British Library Cataloguing in Publication Data
Data available*

Library of Congress Cataloging in Publication Data
Grauwe, Paul de.
The economics of monetary integration/Paul de Grauwe. – 3rd rev. ed.
p. cm.
Includes bibliographical references and index.
1. European Monetary System (Organization) 2. Monetary unions—
European Union countries. 3. Monetary policy—European Union countries.
I. Title.
HG930.5.G674 1997 332.4'5'094—dc21 97–12128
ISBN 0–19–877550–4
ISBN 0–19–877549–0 (Pbk.)

1 3 5 7 9 10 8 6 4 2

Typeset by J&L Composition Ltd, Filey, North Yorkshire
Printed in Great Britain on
acid-free paper by
Bookcraft (Bath) Ltd, Midsomer Norton, Somerset

PREFACE

Just a few years ago academic discussions relating to monetary integration had almost completely vanished. The subject seemed dead. The prospects of a monetary union in Europe, which had triggered earlier discussions, looked very dim. The prevailing academic and non-academic view was that monetary union in Europe was too far away to be worth thinking about.

Things changed very quickly in the middle of the 1980s. A new political momentum was created favouring drastic new initiatives towards economic integration in Europe. Thanks to the Delors Report, which was published in 1989, thinking about monetary integration was made respectable again. Academic studies about the subject became a boom industry. The Maastricht Treaty intensified discussions about why and how to achieve monetary union in Europe.

This book attempts to bring together both the old literature on monetary integration, which was developed in the 1960s and the early 1970s, and the new literature which started to flourish from the middle of the 1980s. As the reader will discover, the confrontation of these two strands of the literature on monetary integration has allowed us to obtain new and exciting insights, which are also important to the policy-maker.

The first part of the book is entirely devoted to what is now called the theory of optimum currency areas. This theory analyses the costs and benefits for nations of joining a monetary union. The basic insights of this theory were developed in the 1960s. Generally speaking this theory led to some scepticism about the benefits of monetary union. The recent literature, however, emphasizing issues of credibility, is much less pessimistic. An attampt is made at integrating these two views. This analysis leads us straight into an issue that has become topical since the Danish and the French referenda about the Maastricht Treaty: should the European Community aim to establish one monetary union encompassing all its members? Or should it rather push for a limited union, allowing some countries to be part of it, while other member-countries stay outside.

In the second part of the book we discuss issues relating to the process that leads to monetary integration. Problems that are studied in this part are the transition to monetary union, the workings of 'incomplete' monetary unions (like the European Monetary System), the setting-up of a common central bank, and fiscal policies in a monetary union. Recent events, following the signing of the Maastricht Treaty, have made these issues of great practical importance. In particular, the fragility of the European Monetary System

(EMS) has appeared to be the weak link in the transition to monetary union. We analyse why the EMS remains a fragile monetary institution. We also study what can be done about it, so that the momentum towards monetary union in Europe is not lost. This analysis will lead us to question the transition strategy towards monetary union as spelt out in the Maastricht Treaty.

This book owes a lot to the criticism and comments of many colleagues who read previous versions of the manuscript. In particular, I am grateful to Filip Abraham, Michael Artis, Bernard Delbecque, Michele Fratianni, Wolfgang Gebauer, Daniel Gros, Ivo Maes, Wim Moesen and Jürgen von Hagen. My gratitude also goes to Gudrun Heyde who patiently checked the manuscript and suggested many improvements in the text. Finally, I am grateful to Francine Duysens for her excellent secretarial help.

Leuven   P. D. G.
October 1992

PREFACE TO THE SECOND EDITION

The ideal book does not require a second edition. It does not need corrections and additions even when the world changes. Unfortunately ours was not an ideal book. The changes that occurred in the European monetary environment called for a second edition.

Fortunately, the corrections and the additions that had to be introduced in the light of the events of the early 1990s were not such that they required a change in the thrust of the arguments developed in the first edition. The predictions contained in the first edition were that the EMS was fragile and would not survive, and that the Maastricht strategy towards monetary union would have to be revised. These predictions have come out.

The particulars of the disintegration of the EMS, however, could not be predicted. Therefore, in this second edition, some importance is attached to the history of the rise and fall of the European Monetary System. In addition, more than in the first edition, there is an analysis of the prospects for monetary union in Europe.

Chapters 5 and 6 which deal with the European Monetary System and with the Maastricht strategy towards monetary union have been rewritten and expanded. In the process of rewriting these chapters we have also been guided by new developments in the academic literature about the credibility of fixed exchange-rate arrangements. In the other chapters, in particular Chapters 1–4 that deal with the theory of optimum currency areas, the changes have been less extensive. Where necessary the empirical evidence was updated, and important new theoretical discussions were added.

In preparing this second edition I have been helped very much by the comments on the first edition made by many colleagues. In particular, my thanks go to Filip Abraham, Juan José Calaza, Harris Dellas, Daniel Gros, Henk Jager, Georgios Karras, and Stefano Micossi. Finally, I am grateful to Nancy Verret for research assistance and to Francine Duysens for secretarial assistance.

Leuven   P. D. G.
January 1994

PREFACE TO THE THIRD EDITION

Since the second edition was published, major decisions have been taken about the transition process towards monetary unification in Europe and about the functioning of the future European monetary institutions. These decisions have affected the prospects of monetary union in Europe. The major remaining uncertainty has to do with the question of which countries will be allowed to join the union in 1999.

All these developments call for a new edition discussing the issues arising from these recent decisions. The first part of the book, analysing the theory of optimum currency areas, has been left relatively unchanged. The only major change is the addition of new empirical evidence concerning the question of whether the European Union constitutes an optimum currency area. In general these new empirical studies tend to be more optimistic about this question than the older empirical literature.

The second part, dealing with the transition towards monetary union and the operation of the future EMU, has been rewritten thoroughly. New issues and problems are discussed, e.g. how to fix the conversion rates at the start of EMU, the stability pact that should govern fiscal policies in the future EMU, the design and the operational procedures of the European central bank, the relations between the 'ins' and the 'outs'.

In preparing this third edition I have benefited a great deal from discussions with colleagues and students. Competent research assistance was provided by Cláudia Costa. I am also grateful to Filip Abraham and Sylvester Eijffinger for their comments, to Yunus Aksoy and Frauke Skudelny for their assistance, and to Francine Duysens for secretarial help.

Leuven   P. D. G.
January 1997

CONTENTS

....

LIST OF FIGURES

LIST OF TABLES

INTRODUCTION

Is there a good *economic* case for countries to have separate currencies? Does a nation increase its welfare when it abolishes its national currency and adopts some currency of a wider area?

Many European citizens today tend to give positive answers to these questions. The mood in Europe towards monetary union is positive. The answer to these questions, however, is not obvious. There are benefits *and* costs to a monetary union.

The question of whether a nation gains by relinquishing its national currency leads immediately to a new question. Suppose Europeans conclude that it would be good economics to eliminate national currencies: Where do they stop then? Should they have one money for the Benelux, or for the EU, or for the whole of Europe, or maybe for the whole world? This problem leads us to raise the question of what constitutes an optimal monetary area.

In order to tackle all these problems we have to analyse systematically what the costs and benefits are of having one currency. In the first part of this book we will analyse these issues. We will study the *economic* costs and benefits exclusively. There is also much to be said about the political costs and benefits. These, however, are outside the scope of this book.

In the first part of the book the focus is on *complete* monetary unions, i.e. a system in which countries abolish their national currencies and substitute a common currency. In the second part we study how *incomplete* monetary unions function. We analyse how monetary systems operate in which national monetary authorities maintain their national currencies but agree to fix their exchange rates (rigidly or less rigidly). This will lead us to an analysis of the European Monetary System. In this second part we also focus on issues relating to the transition to a full monetary union. How should the transition process to a full monetary union be managed? Should this transition be gradual or sudden? What is the nature of the institutions (including the European central bank) that will have to be created? These questions are still hotly debated, and of great importance. The answer to these questions will also determine how successful the future monetary union will be.

PART I
COSTS AND BENEFITS OF MONETARY UNION

1

THE COSTS OF A COMMON CURRENCY

Introduction

The costs of a monetary union derive from the fact that when a country relinquishes its national currency, it also relinquishes an instrument of economic policy, i.e. it loses the ability to conduct a national monetary policy. In other words, in a full monetary union the national central bank either ceases to exist or will have no real power. This implies that a nation joining a monetary union will not be able any more to change the price of its currency (by devaluations and revaluations), or to determine the quantity of the national money in circulation.

One may raise the issue here of what good it does for a nation to be able to conduct an independent monetary policy (including changing the price of its currency). There are many situations in which these policies can be very useful for an individual nation. The use of the exchange rate as a policy instrument, for example, is useful because nations are different in some important senses, requiring changes in the exchange rate to occur. In the next section we first analyse some of these differences that may require exchange rate adjustments. In later sections we analyse how the loss of monetary independence may be costly in some other ways for an individual nation.

The analysis that follows in this chapter is known as the 'theory of optimum currency areas'. This theory which has been pioneered by Mundell (1961), McKinnon (1963), and Kenen (1969) has concentrated on the cost side of the cost-benefit analysis of a monetary union.[1]

1. Shifts in demand (Mundell)

Consider the case of a demand shift developed by Mundell in his celebrated article on optimum currency areas.[2] Let us suppose that for some reason EU

[1] For surveys of this literature, see Ishiyama (1975) and Tower and Willett (1976).
[2] See Mundell (1961).

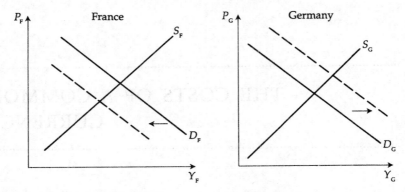

Fig. 1.1. Aggregate demand and supply in France and Germany

consumers shift their preferences away from French-made to German-made products. We present the effects of this shift in aggregate demand in Fig. 1.1.

The curves in Fig. 1.1 are the standard aggregate demand and supply curves in an open economy of most macroeconomics textbooks.[3] The demand curve is the negatively sloped line indicating that when the domestic price level increases the demand for the domestic output declines.[4]

The supply curve expresses the idea that when the price of the domestic output increases, domestic firms will increase their supply, to profit from the higher price. These supply curves therefore assume competition in the output markets. In addition, each supply curve is drawn under the assumption that the nominal wage rate and the price of other inputs (e.g. energy, imported inputs) remain constant. Changes in the prices of these inputs will shift these supply curves.

The demand shift is represented by an upward movement of the demand curve in Germany, and a downward movement in France. The result is that output declines in France and that it increases in Germany. This is most likely to lead to additional unemployment in France and a decline of unemployment in Germany.

The effects on the current accounts of the two countries can also be analysed using Fig. 1.1. We first note that the current account is defined as follows:

current account = domestic output − domestic spending

where these variables are expressed in money terms.

---

[3] See Dornbusch and Fischer (1978) or Parkin and Bade (1988).

[4] This is the substitution effect of a price increase. In the standard aggregate demand analysis, there is also a monetary effect: when the domestic price level increases, the stock of real cash balances declines, leading to an upward movement in the domestic real interest rate. This in turn reduces aggregate demand (see P. De Grauwe, 1983). Here we disregard the monetary effect and concentrate on the substitution effect.

In France the value of domestic output has declined as a result of the shift in aggregate demand. If spending by French residents does not decline by the same amount, France will have a current account deficit. This is the most likely outcome, since the social security system automatically pays unemployment benefits. As a result, the disposable income of French residents does not decline to the same extent as output falls. The counterpart of all this is an increase in the French government budget deficit.

In Germany the situation will be the reverse. The value of output increases. It is most likely that the value of total spending by German residents will not increase to the same extent. Part of the extra disposable income is likely to be saved. As a result, Germany will show a surplus on its current account.

Both countries will have an adjustment problem. France is plagued with unemployment and a current account deficit. Germany experiences a boom which also leads to upward pressures on its price level, and it accumulates current account surpluses. The question that arises is whether there is a mechanism that leads to automatic equilibration, without the countries having to resort to devaluations and revaluations?

The answer is positive. There are two mechanisms that will automatically bring back equilibrium in the two countries. One is based on wage flexibility, the other on the mobility of labour.

(1) Wage flexibility. If wages in France and Germany are flexible the following will happen. French workers who are unemployed will reduce their wage claims. In Germany the excess demand for labour will push up the wage rate. The effect of this adjustment mechanism is shown in Fig. 1.2. The reduction of the wage rate in France shifts the aggregate supply curve downwards, whereas the wage increases in Germany shift the aggregate supply curve upwards. These shifts tend to bring back equilibrium. In France the price of output declines, making French products more competitive, and stimulating demand. The opposite occurs in Germany. This adjustment at the same time

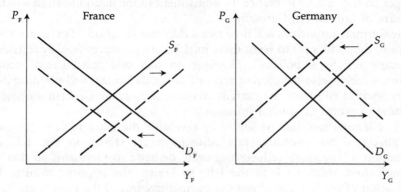

Fig. 1.2. The automatic adjustment process

improves the French current account and reduces the German current account surplus.

Note also that the second-order effects on aggregate demand will reinforce the equilibrating mechanism. The wage and price increases in Germany make French products more competitive. This leads to an upward shift in the French aggregate demand curve. Similarly, the decline in French costs and prices makes German products less competitive and shifts the German aggregate demand curve downwards.

(2) Mobility of labour. A second mechanism that will lead to a new equilibrium involves mobility of labour. The French unemployed workers move to Germany where there is excess demand for labour. This movement of labour eliminates the need to let wages decline in France and increase in Germany. Thus, the French unemployment problem disappears, whereas the inflationary wage pressures in Germany vanish. At the same time the current account disequilibria will also decline. The reason is that the French unemployed were spending on goods and services before without producing anything. This problem tends to disappear with the emigration of the French workers to Germany.

Thus, in principle the adjustment problem for France and Germany will disappear automatically if wages are flexible, and/or if the mobility of labour between the two countries is sufficiently high. If these conditions are not satisfied, however, the adjustment problem will not vanish. Suppose, for example, that wages in France do not decline despite the unemployment situation, and that French workers do not move to Germany. In that case France is stuck in the disequilibrium situation as depicted in Fig. 1.1. In Germany, the excess demand for labour puts upward pressure on the wage rate, producing an upward shift of the supply curve. The adjustment to the disequilibrium must now come exclusively through price increases in Germany. These German price increases make French goods more competitive again, leading an upward shift in the aggregate demand curve in France. Thus, if wages do not decline in France the adjustment to the disequilibrium will take the form of inflation in Germany.

The German authorities will now face a *dilemma* situation. If they care about inflation, they will want to resist these inflationary pressures (e.g. by restrictive monetary and fiscal policies). However, in that case the current account surplus (which is also the current account deficit of France) will not disappear. If they want to eliminate the current account surplus, the German authorities will have to accept the higher inflation.

This dilemma can only be solved by revaluing the mark against the franc. The effects of this exchange rate adjustment are shown in Fig. 1.3. The revaluation of the mark reduces aggregate demand in Germany, so that the demand curve shifts back to the left. In France the opposite occurs. The devaluation of the franc increases the competitiveness of the French products. This shifts the French aggregate demand curve upwards.

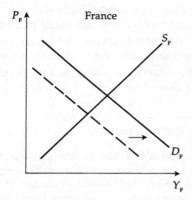
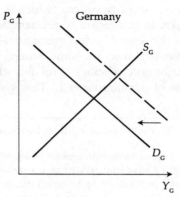

Fig. 1.3. Effects of a revaluation of the DM

The effects of these demand shifts is that France solves its unemployment problem, and that Germany avoids having to accept inflationary pressures. At the same time, the current account deficit of France and surplus of Germany tend to disappear. This remarkable feat is achieved using just one instrument. (The reader may sense that this is too good to be true. And indeed it is. However, for the moment we just present Mundell's theory. We come back later with criticism.)

If France has relinquished the control over its exchange rate by joining a monetary union with Germany, it will be saddled with a sustained unemployment problem, and a current account deficit that can only disappear by deflation in France. In this sense we can say that a monetary union has a cost for France when it is faced with a negative demand shock. Similarly, Germany will find it costly to be in a monetary union with France, because it will have to accept more inflation than it would like.

Can we solve the dilemma in which the two countries find themselves by using other instruments? The answer, in principal, is yes. The German authorities could increase taxes in Germany so as to reduce aggregate demand. (The aggregate demand curve shifts downwards as it does when the mark is revalued.) These tax revenues are then transferred to France where they are spent. (The aggregate demand curve shifts upwards in France.) France would still have a current account deficit. However, it is financed by the transfer from Germany.

It will be obvious that this solution to the problem is difficult to contemplate between sovereign nations, especially since it would have to be repeated every year if the demand shift that started the problem is a permanent one. This solution, however, is frequently applied between regions of the same nation. Many countries have implicit or explicit regional redistribution schemes through the federal budget. *Implicit* regional redistribution within a nation occurs because of the fact that a large part of the government budget is centralized. As a result, when output declines in a region, the tax revenue of the federal government from this region declines. At the same time, however, the social security system (which is quite often also centralized) will increase transfers to this region (e.g. unemployment benefits). The net result of all this is that the central budget automatically redistributes in favour of regions whose income declines. In some federal states there also exist *explicit* regional redistribution schemes. Probably the best known of these is the German system of 'Finanzausgleich'. This system is described in Box 1.

Box 1. Fiscal equalization between Länder in Germany

The system came into existence after the Second World War. Its basic philosophy is that Länder (states) whose tax revenues fall below some predetermined range should receive compensation from Länder whose tax revenues exceed that range. The way this range is computed is rather technical (for more detail see Zimmerman (1989)). Simplifying, it consists in first calculating what the normal tax revenue should be for each state. A state whose tax revenues fall below 92% of this norm receives compensation. To cover these transfers the states whose tax revenues exceed the norm by 2% or more contribute to the system. In Table B1.1 we show the amount of redistribution obtained by this system in 1995.

Table B1.1. Amount of redistribution (in millions of DM) through the system of Finanzausgleich in Germany (1995)

Contributing Länder	
North Rhine–Westphalia	3,442
Baden Württemberg	2,804
Bavaria	2,533
Hessen	2,154
Schleswig–Holstein	142
Hamburg	118
Total	11,193
Receiving Länder	
Berlin	4,209
Saxony	1,783
Saxony–Anhalt	1,123
Thüringen	1,017
Brandenburg	865
Mecklenburg–Vorpommern	771
Bremen	562
Rheinland–Palatinate	229
Lower Saxony	451
Saarland	180
Total	11,190

Source: Bundesministerium der Finanzen (1996), 146.

With German unification the system was expanded to incorporate the new Länder. These have all become net receivers of transfers. This system of redistribution (together with the automatic redistribution resulting from the centralization of the Federal budget) has led to a remarkable reduction of regional income inequalities in Germany.

It should be stressed here that fiscal transfers between regions and countries do not solve the adjustment problem following an asymmetric shock. They just make life easier in the country (region) experiencing a negative demand shock and receiving transfers from the other countries (regions). When the demand shock is a permanent one, price and wage adjustments and/or factor mobility will be necessary to deal with the problem. Fiscal transfers are only suited to dealing with temporary shocks in aggregate demand. If fiscal transfers take on a permanent character, they may even make the adjustment to permanent demand shocks more difficult, as they become a substitute for wage and price changes and for mobility of labour.

Let us recapitulate the main points developed in this section. If wages are rigid and if labour mobility is limited, countries that form a monetary union will find it harder to adjust to demand shifts than countries that have maintained their own national moneys and that can devalue (revalue) their currency. In the latter case, the exchange rate adds some flexibility to a system which is overly rigid. Put differently, a monetary union between two or more countries is optimal if one of the following conditions is satisfied: (*a*) there is sufficient wage flexibility, (*b*) there is sufficient mobility of labour. It also helps to form a monetary union if the budgetary process is sufficiently centralized so that transfers can be organized smoothly (and not after a lot of political bickering) between the countries of the union.

From this analysis one might be tempted to conclude that the conditions for forming a monetary union between, say, France and Germany are most probably not satisfied. This would, however, be too rash a conclusion. We have not yet introduced the benefits of a monetary union. After all, one can only draw conclusions after comparing costs with benefits. In addition, there is some criticism to be levied against the preceding analysis. We will come back to this criticism in Chapter 2.

2. Different preferences of countries about inflation and unemployment

Countries differ also because they have different preferences. Some countries are less allergic to inflation than others. This may make the introduction of a common currency costly. The importance of these differences has been analysed by Corden (1972) and Giersch (1973). We present the problem using a simple graphical representation due to De Grauwe (1975).

Consider two countries. For a change, let us call them Italy and Germany. In Fig. 1.4 we represent the Phillips curves of these two countries, in the right-hand panels. The vertical axis shows the rate of change of the wage rate,

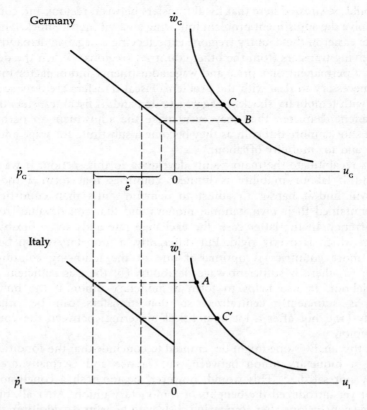

Fig. 1.4. Inflation-unemployment choices in Italy and Germany

the horizontal axis the unemployment rate. (We assume for a moment that these Phillips curves are stable, i.e. they do not shift as a result of changes in expectations of inflation. The modern reader will have difficulties swallowing this. Let him/her be patient. We will ask the question of how the analysis is affected once we take into account the fact that these Phillips curves are not stable.)

In the left-hand side panels we represent the relation between wage changes and price changes. This relationship can be written as follows for Italy and Germany respectively:

$$\dot{p}_I = \dot{w}_I - \dot{q}_I \tag{1.1}$$

$$\dot{p}_G = \dot{w}_G - \dot{q}_G \tag{1.2}$$

where $\dot{p}_I$ and $\dot{p}_G$ are the rates of inflation, $\dot{w}_I$ and $\dot{w}_G$ are the rates of wage increases, and $\dot{q}_I$ and $\dot{q}_G$ are the rates of growth of labour productivity in Italy and Germany. Equations (1.1) and (1.2) can be interpreted by an example.

Suppose wages increase by 10% and the productivity of labour increases by 5% in Italy. Then the rate of price increase that maintains the share of profits in total value added unchanged is 5%. Thus equations (1.1) and (1.2) can be considered to define the rate of price changes that keep profits unchanged (as a percentage of value added).[5] These two equations are represented by the straight lines in the left-hand side panels. Note that the intercept is given by $\dot{q}_I$ and $\dot{q}_G$ respectively. Thus, when the rate of productivity increases in Italy, the line shifts upwards.

The two countries are linked by the purchasing power parity condition, i.e.

$$\dot{e} = \dot{p}_I - \dot{p}_G \tag{1.3}$$

where $\dot{e}$ is the rate of depreciation of the lira relative to the mark. Equation (1.3) should be interpreted as an equilibrium condition. It says that if Italy has a higher rate of inflation than Germany, it will have to depreciate its currency to maintain the competitiveness of its products unchanged. If Italy and Germany decide to form a monetary union, the exchange rate is fixed ($\dot{e} = 0$), so that the rates of inflation must be equal. If this is not the case, e.g. inflation in Italy is higher than in Germany, Italy will increasingly lose competitiveness.

Suppose now that Italy and Germany have different preferences about inflation and unemployment. Italy chooses point A on its Phillips curve, whereas Germany chooses B. It is now immediately obvious that the inflation rates will be different in the two countries, and that a fixed exchange rate will be unsustainable. The cost of a monetary union for the two countries now consists in the fact that if Italy and Germany want to keep the exchange rate fixed they will have to choose another (less preferred) point on their Phillips curves, so that an equal rate of inflation becomes possible. Such an outcome is given by the points C and C' on the respective Phillips curves. (Note that many other points are possible, leading to other joint inflation rates.) Italy now has to accept less inflation and more unemployment than it would do otherwise, Germany has to accept more inflation and less unemployment.

This analysis, which was popular in the 1960s and the early 1970s, has fallen victim to the demise of the Phillips curve. Following the criticisms of Friedman (1968) and Phelps (1968), it is now generally accepted that the Phillips curve is not stable, i.e. that it will shift upwards when expectations of inflation increase. Thus, a country that chooses too high an inflation rate will find that its Phillips curve shifts upwards. Under these conditions the authorities have very little free choice between inflation and unemployment. This has also led to the view that the Phillips curve is really a vertical line in the long run, with far-reaching

---

[5] Suppose a perfect competitive environment. Then profit maximization implies that $w/p = \delta X/\delta L$. If the production function is Cobb-Douglas, $w/p = \alpha X/L$, where α is the labour share in value added. Taking rates of change yields $\dot{w} - \dot{p} = \dot{q}$ (assuming a constant α).

consequences for the costs of a monetary union. We will return to this issue in Chapter 2 when we critically examine the theory of optimum currency areas.

3. Differences in labour market institutions

There is no doubt that there are important institutional differences in the labour markets of European countries. Some labour markets are dominated by highly centralized labour unions (e.g. Germany). In other countries labour unions are decentralized (e.g. the UK). These differences may introduce significant costs for a monetary union. The main reason is that these institutional differences can lead to divergent wage and price developments, even if countries face the same disturbances. For example, when two countries are subjected to the same oil price increase the effect this has on the domestic wages and prices very much depends on how labour unions react to these shocks.

Recent macroeconomic theories have been developed that shed some light on the importance of labour market institutions. The most popular one was developed by Bruno and Sachs (1985).[6] The idea can be formulated as follows. Supply shocks, such as the one that occurred during 1979–80, have very different macroeconomic effects depending upon the degree of centralization of wage bargaining. When wage bargaining is centralized (Bruno and Sachs call countries with centralized wage bargaining 'corporatist'), labour unions take into account the inflationary effect of wage increases. In other words they know that excessive wage claims will lead to more inflation, so that real wages will not increase. They will have no incentive to make these excessive wage claims. Thus, when a supply shock occurs, as in 1979–80, they realize that the loss in real wages due to the supply shocks cannot be compensated by nominal wage increases.

Things are quite different in countries with less centralized wage bargaining. In these countries individual unions that bargain for higher nominal wages know that the effect of these nominal wage increases on the aggregate price level is small, because these unions only represent a small fraction of the labour force. There is a free-riding problem. Each union has an interest in increasing the nominal wage of its members. For if it does not do so, the real wage of its members would decline, given that all the other unions are likely to increase the nominal wage for their members. In equilibrium this non-co-operative game will produce a higher nominal wage level than the co-operative (centralized) game. In countries with decentralized wage bargaining therefore it is structurally more difficult to arrive at wage moderation after a supply shock. In such a non-co-operative set-up no individual union has an incentive to take the first step in reducing its nominal wage claim. For it risks that the others will not follow, so that the real wage level of its members will decline.

---

[6] It should be stressed that these theories were already available and widely discussed in many European countries before Bruno and Sachs discovered them. Their advantage was that they wrote in English, and thereby succeeded in disseminating the idea internationally.

The analogy with the spectators in a football stadium is well known. When they are all seated, the individual spectator has an incentive to stand up so as to have a better view of the game. The dynamics of this game is that they all stand up, see no better, and are more uncomfortable. Once they stand up, it is equally difficult to induce them to sit down. The individual who takes the first step and sits down will see nothing, as long as the others do not follow his example. Since he is sitting, most spectators in the stadium will not even notice this good example.

This co-operation story has been extended by Calmfors and Driffill (1988) who noted that the relationship between centralization of wage bargaining and outcomes is not a linear process. In particular, the more we move towards the decentralized spectrum the more another externality comes to play a role. For in a very decentralized system (e.g. wage bargaining at the firm level), the wage claims will have a direct effect on the competitiveness of the firm, and therefore on the employment prospects of individual union members. Excessive wage claims by an individual union will lead to a strong reduction of employment. Thus, when faced with a supply shock, unions in such a decentralized system may exhibit a considerable degree of wage restraint.

This insight then leads to the conclusion that countries with either strong centralization or strong decentralization of wage bargaining are better equipped to face supply shocks such as the one that occurred during 1979–80 than countries with an intermediate degree of centralization. In these 'extreme' countries there will be a greater wage moderation than in the intermediate countries. As a result, the countries with extreme centralization or decentralization tend to fare better, in terms of inflation and unemployment, following supply shocks, than the others.

Some empirical evidence for this hypothesis is shown in Fig. 1.5. On the horizontal axis we show the degree of centralization of labour markets in a group of industrial countries. (These indices were computed by Calmfors and Driffill (1988).) On the vertical axis the changes in the 'misery' indices of the same countries from the 1970s to the 1980s are represented. These misery indices are the sum of the inflation rate and the unemployment rate.[7] One can see that intermediate countries seem to have experienced a greater worsening of their misery indices than the countries with extreme centralization or decentralization. In other words the labour market institutions of these countries may have made it more difficult to reduce inflation without losses in output following the supply shocks of 1979–80.

It follows that a country might find itself in a situation where wages and prices increase faster than in other countries even when the shock that triggered it all is the same. In terms of the two-country model that we used in Section 1 the supply curve in one country shifts upwards more than in the

[7] In De Grauwe (1990) econometric evidence is presented giving some support to the non-linear relationship between economic performance and the degree of centralization of labour markets.

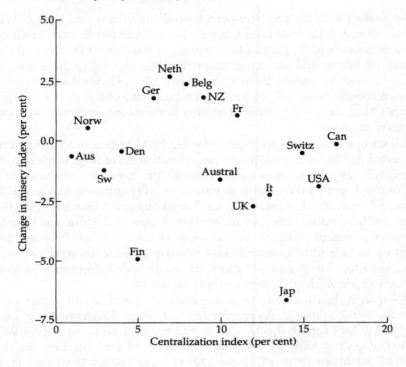

Fig. 1.5. Change in misery indices (from the 1970s to the 1980s) and labour market centralization

Note: The misery index is the sum of the inflation rate and the unemployment rate. The change in the index is measured from 1973–9 to 1980–8. The centralization index measures the degree of centralization of labour unions. A *low* number implies a *high* degree of centralization.

Sources: Misery index computed from OECD, *Economic Outlook*; index of centralization of wage bargaining from Calmfors and Driffill (1988).

other country. This will lead to macroeconomic adjustment problems that are of the same nature as the ones we analysed in Section 1.

We conclude that countries with very different labour market institutions may find it costly to form a monetary union. With each supply shock, wages and prices in these countries may be affected differently, making it difficult to correct for these differences when the exchange rate is irrevocably fixed.

4. Growth rates are different

Some countries grow faster than others. This is made clear in Table 1.1. We find that during the 1980s some southern European countries (and Ireland)

Table 1.1. Average yearly growth rates of
GDP in the EU, 1981–1995

Austria	2.11
Belgium	1.71
Denmark	2.03
Finland	1.88
France	1.95
Germany	2.13
Greece	1.48
Ireland	4.03
Italy	1.85
Netherlands	2.11
Portugal	2.35
Spain	2.47
Sweden	1.41
United Kingdom	2.17

Source: EC Commission, *European Economy*.

experienced growth rates of their GDP which were higher than in the northern part of Europe. (The same phenomenon is observed during the 1970s.)

Such differences in growth rates could lead to a problem when countries form a monetary union. We illustrate it with the following example. Country A's GDP is growing at 5% per year, country B's GDP at 3% per year. Suppose that the income elasticity of A's imports from B is one, and that similarly B's income elasticity of imports from A is equal to one. Then country A's imports from B will grow at 5% per year, whereas B's imports from A will grow at only 3% per year. This will lead to a trade balance problem of the fast-growing country A, whose imports tend to grow faster than its exports.

In order to avoid chronic deficits of its trade account, country A will have to reduce the price of its exports to country B, so that the latter country increases its purchases of goods from country A. In other words, country A's terms of trade must decline so as to make its products more competitive. Country A can do this in two ways: a depreciation of the currency or a lower rate of domestic price increases than in country B. If it joins a monetary union with country B, however, only the second option will be available. This will require country A to follow relatively deflationary policies, which in turn will constrain the growth process. Thus, a monetary union has a cost for the fast-growing country. It will find it more advantageous to keep its national currency, so as to have the option of depreciating its currency when it finds itself constrained by unfavourable developments in its trade account.

5. Different fiscal systems and the seigniorage problem

Countries differ also because they have different fiscal systems. These differences often lead countries to use different combinations of debt and monetary financing of the government budget deficit. When these countries join a monetary union, they will be constrained in the way they finance their budget deficits.

In order to show this, it is useful to start from the government budget constraint:

$$G - T + rB = dB/dt + dM/dt \tag{1.4}$$

where G is the level of government spending (excluding interest payments on the government debt), T is the tax revenue, r is the interest rate on the government debt, B, and M is the level of high-powered money (monetary base).

The left-hand side of equation (1.4) is the government budget deficit. It consists of the primary budget deficit $(G - T)$ and the interest payment on the government debt (rB). The right-hand side is the financing side. The budget deficit can be financed by issuing debt (dB/dt) or by issuing high-powered money dM/dt.

The theory of optimal public finance now tells us that rational governments will use the different sources of revenue so that the marginal cost of raising revenue through these different means is equalized.[8] Thus, if the marginal cost of raising revenue by increasing taxes exceeds the marginal cost of raising revenue by inflation (seigniorage), it will be optimal to reduce taxes and to increase inflation.

The preceding also means that countries will have different optimal inflation rates. In general, countries with an underdeveloped tax system will find it more advantageous to raise revenue by inflation (seigniorage). Put differently, a country with an underdeveloped fiscal system experiences large costs in raising revenue by increasing tax rates. It will be less costly to increase government revenue by inflation.

This reasoning leads to the following implication for the costs of a monetary union. Less developed countries that join a monetary union with more developed countries that have a low rate of inflation will also have to lower inflation. This then means that, for a given level of spending, they will have to increase taxes. There will be a loss of welfare. Some economists (e.g. Dornbusch (1987)) have claimed that this is a particularly acute problem for the southern EC countries. By joining the low-inflation northern monetary zone they will have to increase taxes, or let the deficit increase further. For these countries the cost of the monetary union is that they will have to rely too much on a costly way of raising revenues.

Table 1.2 gives some empirical evidence on the size of the seigniorage for these southern countries, and compares it with Germany. We observe that up to the middle of the 1980s the southern European countries had high seigniorage revenues. These revenues amounted to 2–3% of GNP in all these countries. This was certainly much more important than in the northern countries. Since the middle of the 1980s, however, seigniorage revenue of these countries

---

[8] See Fischer (1982) and Grilli (1989).

Table 1.2. Seigniorage revenues as per cent of GNP

	1976–85	1986–90	1993
Germany	0.2	0.6	0.5
Greece	3.4	1.5	0.7
Italy	2.6	0.7	0.5
Portugal	3.4	1.9	0.6
Spain	2.9	0.8	0.6

Sources: Dornbusch (1987); Gros (1990); Gros and Thygesen (1992).

has declined significantly, mainly because of the reduction in their inflation rates. This leads to the conclusion that if these countries should join a monetary union with the low-inflation countries in the 1990s, the additional cost (in terms of public finance) would probably not be very important.

6. Conclusion

In this chapter we discussed differences between countries. We observed that countries can use exchange rate changes, or other monetary policies, to correct for these differences. We found that in most cases there is an alternative to using the exchange rate as a policy instrument. For example, when confronted with a loss of domestic competitiveness, countries can use contractionary demand policies aiming at regaining competitiveness. However, these alternatives are often more painful, and therefore less desirable. To the extent that these alternative policies are more painful than changing the exchange rate we concluded that the country under consideration does not gain from relinquishing its money and joining a currency union. (Note, however, that we still have not introduced the benefit side of the analysis. It is still possible that even if there are costs associated with relinguishing one's national money, the benefits outweigh these costs.)

The analysis of this chapter which is based on the theory of optimum currency areas has recently been subjected to much criticism. This has led to new and important insights. In the next chapter we turn our attention to this criticism.

2

THE THEORY OF OPTIMUM CURRENCY AREAS: A CRITIQUE

Introduction

In the previous chapter we analysed the reasons why countries might find it costly to join a monetary union. Recently, this analysis, which is known as the theory of optimum currency areas, has come under criticism.[1] This criticism has been formulated at different levels. First one may question the view that the differences between countries are important enough to bother about. Secondly, the exchange rate instrument may not be very effective in correcting for the differences between nations. Thirdly, not only may the exchange rate be ineffective, it may do more harm than good in the hands of politicians.

In this chapter we analyse this criticism in greater detail.

1. How relevant are the differences between countries?

There is no doubt that countries *are* different. The question, however, is whether these differences are important enough to represent a stumbling-block for monetary unification.

1.1. Is a demand shock concentrated in one country a likely event?

The classical analysis of Mundell started from the scenario in which a demand shift occurs away from the products of one country in favour of another country. Is such a shock likely to occur frequently between the European countries that are planning to form a monetary union? Two views have emerged to answer this question. We will call the first one the European

[1] See EC Commission (1990) and Gros and Thygesen (1991).

Commission view, which was defended in the report 'One Market, One Money'. The second view is associated with Paul Krugman.

According to the European Commission, differential shocks in demand will occur less frequently in a future monetary union. The reason is the following. Trade between the industrial European nations is to a large degree intra-industry trade. The trade is based on the existence of economies of scale and imperfect competition (product differentiation). It leads to a structure of trade in which countries buy and sell to each other the same categories of products. Thus, France sells cars to and buys cars from Germany. And so does Germany. This structure of trade leads to a situation where most demand shocks will affect these countries in a similar way. For example, when consumers reduce their demand for cars, they will buy fewer French *and* German cars. Thus, both countries' aggregate demand will be affected in similar ways.

The removal of barriers with the completion of the single market will reinforce these tendencies. As a result, most demand shocks will tend to have similar effects.[2] Instead of being asymmetric, these shocks will tend to be more symmetric.

The second and opposite view has been defended by Paul Krugman. According to Krugman (1991), one cannot discard Mundell's analysis, for there is another feature of the dynamics of trade with economies of scale that may make Mundell's analysis very relevant. Trade integration which occurs as a result of economies of scale also leads to regional concentration of industrial activities.[3] The basic argument here is that when impediments to trade decline this has two opposing effects on the localization of industries. It makes it possible to produce closer to the final markets, but it also makes it possible to concentrate production so as to profit from economies of scale (both static and dynamic). This explains why trade integration in fact may lead to more concentration of regional activities rather than less.

The fact that trade may lead to regional concentration of industrial activities is illustrated rather dramatically by comparing the regional distribution of the automobile production in the USA and in Europe (see Table 2.1). The most striking feature of this table is that the US production of automobiles is much more regionally concentrated than the EU's. (This feature is found in many other industrial sectors; see Krugman (1991). There is also no doubt that the US market is more highly integrated than the EU market, i.e. there are fewer impediments to trade in the USA than in the EU. This evidence therefore suggests that when the EU moves forward in the direction of a truly integrated market, it may experience similar kinds of regional concentrations of economic activities as those observed in the USA today. It is therefore not to be excluded that the automobile industry, for example, will tend to be more concentrated

---

[2] Peter Kenen (1969) also stressed the importance of the similarity of the trading structure for making a monetary union less costly.

[3] This is an old idea that was developed by Myrdal (1957) and Kaldor (1966). For a survey see Balassa (1961). Krugman gives a more rigorous underpinning of these ideas in some of his recent writing. See Krugman (1991).

Table 2.1. Distribution of auto production

	USA		EU
Midwest	66.3	Germany	38.5
South	25.4	France	31.1
West	5.1	Italy	17.6
North-east	3.32	UK	12.9

Source: Krugman (1991).

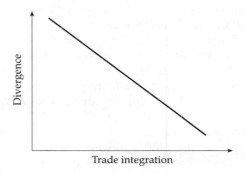

Fig. 2.1. The European Commission view

in, say, Germany (although we are not sure it will be Germany, it could also be another country). Sector-specific shocks may then become country-specific shocks. Countries faced with these shocks may then prefer to use the exchange rate as an instrument of economic policy to correct for these disturbances.

The two views about the relation between economic integration and the occurrence of asymmetric shocks can be analysed more systematically using the graphical device represented in Fig. 2.1. Let us first represent the European Commission view. On the vertical axis we set out the degree of divergent movements of output and employment between groups of countries (regions) which are candidates to form a monetary union.[4] On the horizontal axis we set out a measure of the degree of trade integration between these countries. This measure could be the mutual trade of these countries as a share of their GDP. The European Commission view can then be represented by a downward sloping line. It says that as the degree of economic integration between countries increases, asymmetric shocks will occur less frequently (so that income and employment will tend to diverge less between the countries involved).

---

[4] We could take as the measure of divergence one minus the correlation coefficient between the growth rates of output of these countries. Thus when the correlation is one, our measure of divergence is zero. When the correlation is −1 our measure of divergence would be 2, its maximum value.

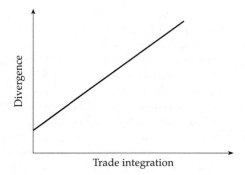

Fig. 2.2. The Krugman view

We represent the second view, which we label the Krugman view, in Fig. 2.2. Instead of a downward sloping line we have a positively sloped line. Thus, when economic integration increases, the countries involved become more specialized so that they will be subjected to more rather than fewer asymmetric shocks.[5]

What is the right view of the world? A clear-cut answer will be difficult to formulate. Nevertheless it is reasonable to claim that a presumption exists in favour of the European Commission view. The reason can be formulated as follows. The fact that economic integration can lead to concentration and agglomeration effects cannot be disputed. At the same time, however, it is also true that as market integration between countries proceeds, national borders become less and less important as factors that decide the location of economic activities. As a result, it becomes more and more likely that concentration and agglomeration effects will be blind to the existence of borders. This creates the possibility that the clusters of economic activity will encompass borders. Put differently, it becomes more and more likely that the relevant regions in which some activity is centralized will transgress one or more borders. For example, it could very well be that automobile manufacturing will not be centralized in Germany, but rather in the region encompassing South Germany and Northern Italy. If this is the case, shocks in the automobile industry will affect Germany *and* Italy, so that the DM–lira rate cannot be used to absorb this shock.

Note that the argument we develop here is not that integration does not lead to concentration effects (it probably will), but rather that national borders will

[5] This view could be associated with Kenen (1969), who stressed that countries with a less diversified output structure are subject to more asymmetric shocks, making them less suitable to form a monetary union. The presumption is that small countries which are highly integrated with the rest of the world are also highly specialized. This leads to the paradox that small and very open countries should keep their own currencies and not join a monetary union (see Frankel and Rose (1996) on this paradox and how it can be resolved).

increasingly be less relevant in influencing the shape of these concentration effects. As a result, regions may still be very much affected by asymmetric shocks. The probability that these regions overlap existing borders, however, will increase as integration moves on. We conclude that the economic forces of integration are likely to rob the exchange rates between national currencies of their capacity to deal with these shocks.

From the preceding arguments it should not be concluded that economists know for sure what the relationship is between economic integration and the occurrence of asymmetric shocks. All we can say is that there is a theoretical presumption in favour of the hypothesis that integration will make asymmetric shocks between nations less likely. The issue remains essentially an empirical one. Recently Frankel and Rose (1996) have undertaken important empirical research relating to this issue. They analysed the degree to which economic activity between pairs of countries is correlated as a function of the intensity of their trade links. Their conclusion was that a closer trade linkage between two countries is strongly and consistently associated with more tightly correlated economic activity between the two countries. In terms of Figs. 2.1 and 2.2 this means that the relationship between divergence and trade integration is negatively sloped.

Similar evidence is presented in Artis and Zhang (1995), who find that as the European countries have become more integrated during the 1980s and 1990s, the business cycles of these counries have become more correlated.

1.2 Asymmetric shocks and the nation-state

In the preceding section we argued that economic integration will tend to reduce the probability that individual nations (in contrast to regions) will be hit by asymmetric shocks. This does not mean, of course, that all asymmetric shocks will disappear. There is another source of asymmetric shocks that will continue to play a role in the future. This is due to the continued existence of nation-states as the main instruments of economic policies.

In the future European monetary union, monetary policies will be centralized, and will therefore cease to be a source of asymmetric shocks. The member countries of the monetary union, however, will continue to exercise considerable sovereignty in a number of economic areas. The most important one is the budgetary field. In the future monetary union most of the spending and taxing powers will continue to be vested in the hands of national authorities. Today, in most EU countries spending and taxation by the national authorities amount to close to 50% of GDP. The spending and taxing powers of the European authorities represent less than 1.5% of GDP. This situation will be maintained after the start of monetary union in 1999. By changing taxes and spending the authorities of an individual country can create large asymmetric shocks. By their very nature these shocks are well contained within national borders. For example, when the authorities of a country increase taxes on wage income, this only affects labour in that country and will influence spending and wage levels in that country. As a result, the aggregate demand and supply curves of the country involved will shift, creating disturbances that will lead to divergent price and wage developments. We are back in the Mundell analysis (of Chapter 1, Section 1) about how to adjust to these asymmetric shocks.

The fact that countries will maintain most of their budgetary powers in a future monetary union creates the possibility that large asymmetric shocks may occur in the union. This raises the issue of how budgetary policies should be conducted in a monetary union. We will return to this issue in a separate chapter (see Chapter 8).

There are other aspects of the existence of nation-states that can be a source of asymmetric disturbances. Many economic institutions are national. The best example is the wage bargaining process, which differs widely between countries, and which can create asymmetric disturbances. We discussed some of the issues relating to the difference in national wage bargaining systems in Section 3 of the previous chapter. In the next section we return to some of these issues.

The previous discussion leads to the conclusion that although economic integration is likely to weaken the occurrence of asymmetric shocks, the existence of nation-states with their own peculiarities will be a continued source of asymmetric disturbances in a monetary union, creating the problems of adjustment we discussed in Chapter 1. This has led some economists to argue that a monetary union can only function satisfactorily if further steps towards political unification are taken. In the view of these economists, the absence of a political union will create great risks of difficult adjustments to (political) disturbances in a future monetary union. It is equally possible, however, that the existence of a monetary union will exert sufficient pressure on the member countries to accelerate their efforts towards establishing a political union. In this view, monetary union will work as a device forcing European countries towards political union. Some of the issues relating to the link between monetary union and political union will be taken up again in Chapter 8 where we discuss budgetary policies in a monetary union.

1.3. Institutional differences in labour markets

The differences in the workings of the labour markets in different countries are well documented. These differences, however, have accumulated over the years, partly because European countries have experienced separate policy regimes. The issue is whether monetary integration will not drastically change the behaviour of labour unions, so that the differences we observe today may disappear.[6]

An example may clarify this point. In Fig. 2.3 we present the labour markets of two countries that are candidates for a monetary union. The figure is based

---

[6] See Gros and Thygesen (1991), ch. 9.

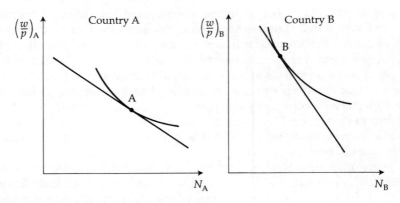

Fig. 2.3. The Solow-McDonald model in two countries

on the model of McDonald and Solow (1981).[7] On the vertical axis we have the real wage level, on the horizontal axis the level of employment (N). The convex curves are the indifference curves of the labour union. It is assumed that there is only one labour union in each country. The union maximizes its utility which depends on both the real wage level and the employment of its members. The negatively sloped line is the economy-wide demand-for-labour curve. For the union, which maximizes its utility, the demand-for-labour curve is a constraint: thus, the union will select a point on it which maximizes its utility. This is represented in Fig. 2.3 by the points A and B.

The interesting feature of this model is that the employment line takes into account the reaction of the authorities to what the labour unions are doing. If we assume that the authorities give a higher weight to employment in their utility function than the labour unions, we may have the following situation. When the labour unions set a wage that reduces the employment level below the level that the authorities find optimal, they will react by changing their policies. For example, they will engage in more expansionary monetary and fiscal policies to absorb the unemployed, they may create public jobs, etc. To the extent that labour unions take this reaction of the authorities into account, the constraint the unions face will change. More specifically, the employment line becomes steeper because an increase in the real wage level reduces private employment and, thus, induces the authorities to intensify their job-creating policies. As a result, an increase in the real wage level has a less pronounced effect on the total level of employment.[8] Thus, the steepness of this employment line also reflects the willingness of the authorities to engage in expansionary employment policies when the wage rate increases.

---

[7] For a discussion of this model see Carlin and Soskice (1990).

[8] This employment line must in fact be interpreted as the reaction curve of the government. The union operates as a 'Stackelberg' leader and selects the optimal point on this reaction line.

In Fig. 2.3 we have drawn the employment line of country B steeper than that of country A, assuming that the authorities of country B are more willing to accommodate the unions' wage-setting behaviour by expansionary employment policies. Monetary union now will change the possibility for the national governments to follow such accommodating policies. Monetary policies will now be centralized, so that the unions of the two countries face the same reactions of the monetary authorities. This will make the employment lines similar, so that the two unions will tend to select a similar combination of wage rates and employment levels.

The differences, however, are unlikely to disappear completely. National governments have other employment policies at their disposal besides monetary policies. For example, they can create jobs in the government sector, financing these extra expenditures by issuing debt. A monetary union does not necessarily constrain this accommodating government behaviour. Thus, although the differences in the behaviour of the labour unions will be less pronounced, they will certainly not be eliminated completely.

Finally, it should also be stressed that the previous analysis assumes a completely centralized union in both countries. As pointed out earlier, unions are different across countries because of different degrees of centralization. It is not clear how monetary union will change these institutional differences.

We conclude that the institutional differences in the national labour markets will continue to exist for quite some time after the introduction of a common currency. This may lead to divergent wage and employment tendencies, and severe adjustment problems when the exchange rate instrument has disappeared.

1.4 Do differences in growth rates matter?

Fast-growing countries experience fast-growing imports. In order to allow exports to increase at the same rate, these countries will have to make their exports more competitive by real depreciations of their currencies. If they join a monetary union, this will be made more difficult. As a result, these countries will be constrained in their growth. This popular view of the constraint imposed on fast-growing countries that decide to join a monetary union has very little empirical support.

In Fig. 2.4 we present data on the growth rates of EC countries during 1976–95 and their real depreciations (or appreciations). The fast-growers are above the horizontal line, the slow-growers below. We observe that among the fast-growers there are countries that saw their currency appreciate and others depreciate. The same is true for the slow-growers.[9]

[9] This visual evidence is confirmed by the regression equation

$$GDP = 0.0 - 0.12^*REER \qquad \text{Corrected } R^2 = 0.01$$
$$\quad (0.0) \quad (0.11)$$

where GDP = growth rate of GDP minus EC growth, and REER = the average growth rate of unit labour costs relative to Community partners. Standard errors are in brackets. It can be seen that the coefficient of REER is not statistically different from zero.

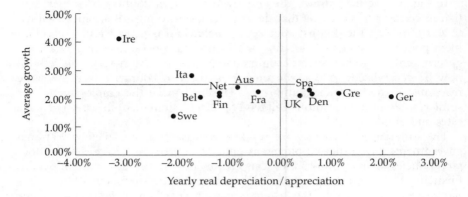

Fig. 2.4. Real depreciation and growth, 1976–1995

Note: Negative values indicate real depreciation, whereas positive values indicate appreciation.

Source: Computed from International Monetary Fund, *IFS*.

This lack of relation between economic growth and real depreciations has been given an elegant interpretation by Paul Krugman (1989). Economic growth has relatively little to do with the static view implicit in the story told in the earlier sections. Economic growth implies mostly the development of new products. Fast-growing countries are those that are able to develop new products, or old products with new qualitative features. The result of this growth process is that the income elasticities of the exports of fast-growing countries are typically higher than those of slow-growers. More importantly, these income elasticities of the export goods of the fast-growers will also typically be higher than the income elasticities of their imports. (See Krugman (1989) for empirical evidence.) As a result, these countries can grow faster without incurring trade balance problems. This also implies that the fast-growers can increase their exports at a fast pace without having to resort to real depreciations.

There is a second reason why the fast-growing countries should not worry too much that joining a monetary union will constrain their potential for growth. This has to do with the existence of capital flows. A fast-growing country is usually also a country where the productivity of capital is higher than in slow-growing countries. This difference in the productivity of capital will induce investment flows from the slow-growing countries to the fast-growing countries. These capital flows then make it possible for the fast-growing country to finance current account deficits without any need to devalue the currency.

There is even an argument to be made here that fast-growing countries that join a monetary union with slow-growing countries will find it easier to attract foreign capital. With no exchange rate uncertainty, investors from the slow-growing area may be more forthcoming in moving their capital to the fast-growing country in order to profit from the larger returns.

One can conclude that differences in the growth rates of countries cannot really be considered as an obstacle to monetary integration. In other words, fast-growing countries will, in general, not have to reduce their growth rates by joining a monetary union.

2. Nominal and real depreciations of the currency

The cost of relinquishing one's national currency lies in the fact that a country cannot change its exchange rate any more to correct for differential developments in demand, or in costs and prices. The question, however, is whether these exchange rate changes are effective in making such corrections. Put differently, the question that arises is whether *nominal* exchange rate changes can permanently alter the *real* exchange rates of the country.

This is a crucial question. For if the answer is negative, one can conclude that countries, even if they develop important differences between themselves, would not have to meet extra costs when joining a monetary union. The instrument they lose does not really allow them to correct for these differences.

In order to analyse this question of the effectiveness of exchange rate changes we return to two of the asymmetric disturbances analysed in the previous chapter.

2.1. Devaluations to correct for asymmetric demand shocks

Let us take the case of France developed in section 1.1 of the previous chapter. We assumed that a shift occurred away from French products in favour of German products. In order to cope with this problem France devalues its currency. We present the situation in Fig. 2.5. As a result of the devaluation aggregate demand in France shifts back upwards and corrects for the initial unfavourable demand shift. The new equilibrium point is F.

It is unlikely that this new equilibrium point can be sustained. The reason is that the devaluation raises the price of imported goods. This raises the cost of production directly. It also will increase the nominal wage level in France as workers are likely to be compensated for the loss of purchasing power. All this means that the aggregate supply curve will shift upwards. Thus, prices increase and output declines. These price increases feed back again into the wage-formation process and lead to further upward movements of the aggregate supply curve. The final equilibrium will be located at a point like F'. The initial favourable effects of the devaluation tend to disappear over time. It is not

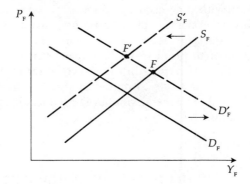

Fig. 2.5. Price and cost effects of a devaluation

possible to say here whether these favourable effects of the devaluation on output will disappear completely. This depends on the openness of the economy, on the degree to which wage-earners will adjust their wage claims to correct for the loss of purchasing power. There is a lot of empirical evidence, however, that for most of the European countries this withering away of the initially favourable effects of a devaluation will be strong.[10]

The previous conclusion can also be phrased as follows. Nominal exchange rate changes have only temporary effects on the competitiveness of countries. Over time the nominal devaluation leads to domestic cost and price increases which tend to restore the initial competitiveness. In other words, nominal devaluations only lead to temporary *real* devaluations. In the long run nominal exchange rate changes do not affect the real exchange rate of a country.

Does this conclusion about the long-run ineffectiveness of exchange rate changes imply that countries do not lose anything by relinquishing this instrument? The answer is negative. We also have to analyse the short-term effects of an exchange rate policy aiming at correcting the initial disturbance, and we have to compare these to alternative policies that will have to be followed in the absence of a devaluation. This is done in Fig. 2.6. We have added here a line (TT) which expresses the trade account equilibrium condition. It is derived as follows. Trade account equilibrium is defined as equality between the value of domestic output and the value of spending by residents (sometimes also called absorption). Thus we have equilibrium in the trade account if and only if:

$$P_d Y = P_a A \tag{2.1}$$

where P_d is the price of the domestic good, Y the domestic output level, P_a is the average price index of the domestic and the imported good, A is absorption (in real terms). The level of real absorption depends on many factors (e.g.

---

[10] See EC Commission (1990), ch. 6.

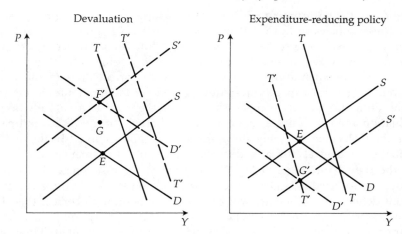

Fig. 2.6. Devaluation and deflationary policies compared

government spending, the real interest rate). If these are fixed, we can derive a negative relation between P_d and Y which maintains the equality (2.1), in other words which maintains trade account equilibrium. The negative relationship follows from the fact that as P_d increases, P_a (which contains the import price) increases less than proportionately, so that the left-hand side increases relative to the right-hand side of (2.1). Thus, when P_d increases, the value of output increases relative to the value of absorption, tending to produce a trade account surplus. It follows that domestic output should decline to maintain trade account equilibrium. Put differently an increase in P_d (for a given import price) is equivalent to an improvement in the terms of trade. This allows the country to reduce domestic output and still maintain equilibrium in the trade account.

Points to the left of this *TT*-line are ones where the country has a trade deficit, i.e. the level of domestic output is too low compared to the level of absorption. Points to the right of the *TT*-line are those where the country produces more than it spends.

In Fig. 2.6, we assume that the country has been hit by a negative demand shock which has brought the output point to *E*. As a result, the country has a trade account deficit which will have to be corrected. One way to correct the disequilibrium is to devalue the currency. The dynamics of the adjustment after the devaluation is shown in the left-hand panel of Fig. 2.6. The devaluation shifts the demand and the supply curves upwards, as explained earlier. However, it also shifts the *TT*-curve to the right (to *TT′*). This follows from the fact that the devaluation increases the price of imports. Thus, in equation (2.1), P_a increases. It follows that P_d and/or Y must increase to maintain trade account equilibrium (at least if the real level of absorption A remains unchanged).

The new equilibrium in the goods market is now located at point F'. It can be seen that there is still a trade account deficit because the new output point is located to the left of the *TT′* line. In order to restore trade account equilibrium, the government will have to follow policies reducing real absorption. These policies have the effect of shifting the TT'-line to the left. The reason is that with a lower level of absorption (due to say lower government spending) the level of output that maintains trade account equilibrium is also reduced. These expenditure-reducing policies also affect the demand curve, however. In general expenditure-reducing policies reduce domestic demand for the domestic good. It can be shown that there exists a combination of devaluation and expenditure-reducing policy that will bring the economy to a point like *G*, located vertically above the initial output point.[11] In the end the devaluation is neutral in that it does not affect output permanently.

In the right-hand panel of Fig. 2.6 we present the case where the country chooses *not* to devalue. Since in the initial situation (point *E*) there is a trade account deficit, the authorities will have to do something to correct this. This will necessarily have to be a policy which reduces absorption. Thus, deflationary monetary and/or fiscal policies will have to be instituted. These shift the trade account equilibrium line *TT* to the left.

These expenditure-reducing policies, however, also reduce aggregate demand for the domestic goods. Thus, the aggregate demand line also shifts to the left. The economy will go through a deflationary process, which reduces output. With sufficient wage and price flexibility, this will also tend to shift the supply line downwards, because the decline in prices leads to lower nominal wages. If wages and prices are not very flexible, this may require a considerable time. In the long run, the economy will settle at a point like *G'*. The output level is equal to its initial level, and the trade account is in equilibrium.

We conclude that in *the long run* the two policies (devaluation and expenditure-reducing) lead to the same effect on output and the trade account. Put differently, in the long run the exchange rate will not solve problems that arise from differences between countries that originate in the goods markets. This result is also in the tradition of the classical economists. These stressed that money is a veil. Structural differences should be tackled with structural policies. Manipulating money (its quantity or its price) cannot change these real differences.

The difference between the two policies, a devaluation or an expenditure-reducing policy, is to be found in their *short-term dynamics*. When the country devalues, it avoids the severe deflationary effects on domestic output during the transition. The cost of this policy is that there will be inflation. With the second policy, inflation is avoided. The cost, however, is that output declines during the transition period. In addition, as we have seen, this second policy may take a long time to be successful if the degree of wage and price flexibility is limited.

One can conclude that although a devaluation does not have a permanent effect on competitiveness and output, its dynamics will be quite different from the dynamics engendered by the alternative policy which will necessarily have to be followed if the country has relinquished control over its national money. This loss of a policy instrument will be a cost of the monetary union.

---

[11] See De Grauwe (1983), ch. 9.

In Box 2 we present a case-study of a devaluation (Belgium in 1982) that helped this country to restore domestic and trade account equilibrium at a cost that was most probably lower than if it had not used the exchange rate instrument. There were other noteworthy and successful devaluations during the 1980s. The French devaluations of 1982–3 (coming after a period of major policy errors) stand out as success stories (see Sachs and Wyplosz (1986)). Similarly, the Danish devaluation of 1982 was quite successful in re-establishing external equilibrium without significant costs in terms of unemployment (see De Grauwe and Vanhaverbeke (1990)).

Box 2. The devaluation of the Belgian franc of 1982

In 1982 Belgium devalued its currency by 8.5%. In addition, fiscal and monetary policies were tightened, and an incomes policy, including temporary abolition of the wage-indexing mechanism, was instituted. This decision came after a period of several speculative crises during which the BF was put under severe pressures. These crises were themselves triggered by the increasing loss of competitiveness (in turn due to excessive wage increases) which the Belgian economy experienced during the 1970s. This has led to unsustainable current account deficits.

It can now be said that the devaluation (together with the other policy measures) was a great success. Not only did it lead to a rapid turnaround in the current account of the balance of payments (see Table B2.1). It managed to do so without imposing great deflationary pressures on the Belgian economy. As Table B.2.2 shows, there was a pronounced recovery in employment after 1983. This recovery in employment proceeded at a pace that was not significantly different from the rest of the Community after that date.

Table B2.1 Current account of Belgium (as a per cent of GDP)

1981	−3.8
1982	−3.6
1983	−0.6
1984	−0.4
1985	0.5
1986	2.0

Source: EC Commission (1990).

Table B2.2 Growth rate of employment (Belgium and EC)

	Belgium	EC
1981	−2.0	−1.2
1982	−1.3	−0.9
1983	−1.1	−0.7
1984	0.0	0.1
1985	0.8	0.6
1986	1.0	0.8

Source: EC Commission (1990).

2.2 Devaluations to correct for different policy preferences

In the previous chapter we presented a model of two countries in which the authorities have different preferences concerning the choice between inflation and unemployment. It was argued that by making suitable exchange rate adjustments countries could choose a preferred point on their Phillips curve. Italy, for example, could by a policy of continuous depreciations of the lira ensure that it could 'buy' a low unemployment rate by accepting more inflation. The cost for Italy of joining the monetary union is that it would have to accept more unemployment than it desired in exchange for less inflation than it desired.

We have already mentioned that this analysis depends on the assumption that the Phillips curve is a stable one, and does not move with changes in expected inflation. The monetarist critique of the Phillips curve has changed all this, and therefore also the analysis of the costs of a monetary union.

The core of the monetarist critique is that a country which chooses too high an inflation rate (and in the process is forced to let its currency depreciate) will find that its Phillips curve shifts upwards. In this monetarist view of the world, the Phillips curve is really a vertical line in the long run. The implications for the costs of a monetary union are analysed in Fig. 2.7, where we represent both the short-term and the long-term (vertical) Phillips curves. The intercept of the long-run vertical Phillips curve with the x axis represents the 'natural' rate or unemployment. We now observe that, in the long run, the authorities cannot choose an optimal combination of inflation and unemployment. The latter is determined by the natural rate of unemployment and is independent of inflation.

There is therefore also nothing to be gained by Italy and Germany from having two different inflation rates. They can set their inflation rates equal to each other by fixing their exchange rates, without any cost in terms of unemployment. Italy and Germany can form a monetary union without costs. Put differently, the fact that Italy and Germany cannot follow independent monetary policies in a monetary union is no loss at all, since an independent monetary policy (and therefore inflation rate) does not bring about lower unemployment.

This analysis is now generally accepted. There remains the problem, however, of the short-term costs of joining a monetary union. Although in the long run, countries cannot really choose between inflation and unemployment, the short-run Phillips curve is still alive. That is, countries that want to reduce inflation will most probably be faced with a temporary increase in the unemployment rate. The experiences of the 1980s makes this clear. In Box 2 we present a few case-studies that illustrate how policies of disinflation during the early 1980s led to significant increases in unemployment in major industrial countries.

The problem that arises then is whether the decision to join a monetary union by a country with a high rate of inflation (Italy in our example of Fig.

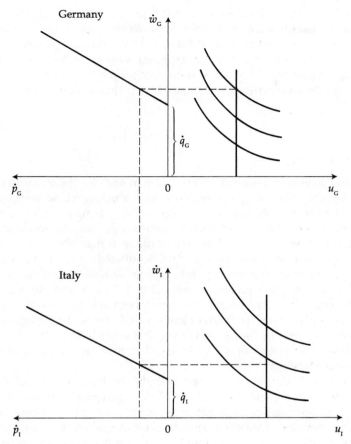

Fig. 2.7. Monetary union in a world of vertical Phillips curves

1.4) may not lead to a temporary but significant unemployment cost. In Fig. 1.4, the decision by Italy to form a monetary union with Germany raises unemployment in Italy from A to C'. Over time the Italian Phillips curve will shift downwards because of lower expectations of inflation. However, during the transition, Italy is faced with high unemployment.

It should be stressed that this cost of disinflating the economy should not necessarily be called a cost of monetary union. If Italy has too high an inflation, it will have to take action to reduce it. It will then face the short-term unemployment cost whether it is part of a monetary union or not. The issue then really boils down to the question of whether it will be less costly for Italy to reduce the rate of inflation when it forms a monetary union with Germany than when it does it alone. This question has been discussed in great detail recently. We shall return to it in the next section, where we discuss issues of the credibility of monetary policies.

The model of Fig. 2.7 allows us to highlight another important source of possible differences between countries. Suppose the growth rate of productivity $\dot{q}$ is higher in Germany than in Italy. If both countries decide to form a monetary union this implies that the nominal wage increases in Italy must be lower than in Germany. This can be seen by setting $\dot{e} = 0$ in equation (1.3), so that $\dot{p}_G = \dot{p}_I$. From equations (1.1) and (1.2) it then follows that

$$\dot{w}_G - \dot{q}_G = \dot{w}_I - \dot{q}_I \tag{2.2}$$

or

$$\dot{w}_G - \dot{w}_I = \dot{q}_G - \dot{q}_I. \tag{2.3}$$

If, following monetary unification, the German and the Italian labour unions should centralize their wage bargaining and aim for equal nominal wage increases despite the differences in productivity growth, this would spell trouble. Italian industry would become increasingly less competitive. Therefore, a condition for a successful monetary union is that labour unions should *not* centralize their wage bargaining when productivity growth rates differ.

We can summarize the main points of this section as follows. Countries differ in terms of their preferences towards inflation and unemployment. These differences, however, cannot be a serious obstacle to forming a monetary union, if one accepts that countries cannot really choose an optimal point on their Phillips curves. The only serious issue that arises in this connection is that (high-inflation) countries that join in a union may face a transitory cost in terms of unemployment.

To conclude, it should be stressed that the analysis of the Phillips-curve model in this section and the analysis of the aggregate demand and supply model in the previous section are very similar. In both sections we have stressed that policies of inflation and devaluations of the currency have only temporary effects on output and employment. In the long run, inflation and depreciations of the currency have no or only limited effects on these variables. In the monetarist models of the world (e.g. a vertical long-run Phillips curve, or a vertical long-run aggregate supply curve) inflation and depreciations of the currency have no effects on output and employment. As a result, in this monetarist world the cost of a monetary union is really zero.

Box 3. The cost of disinflation: some evidence from the 1980s

Fig. B3.1a–f presents data on inflation and unemployment in the major industrialized countries during the 1980s. We observe that in all countries the supply shocks of the 1979–80 period led to an increase in the inflation rate. Most countries then started a process of disinflation, which contributed to an increase in the unemployment rate. Some countries (France, Germany, and Italy) experienced significantly more difficulties in reducing their inflation rates, that is, the cost in terms of unemployment seems to have been considerable. In other words it took many years for the respective Phillips curves to move inwards. For France and Italy the inward movements of the Phillips curves at the end of the 1980s were very small. This suggests two possible interpretations. One is that inflationary expectations were very slow to come down; the other is that the natural rate of unemployment may have increased in these countries.

Countries like the USA and the UK experienced less difficulty in shifting their Phillips curves downward. In De Grauwe (1990) an interpretation is given of this phenomenon. We return to this issue when we discuss the functioning of the European Monetary System in Chapter 5.

Finally, note the special position of Japan, which seems to have been able to reduce inflation without any apparent cost in terms of unemployment.

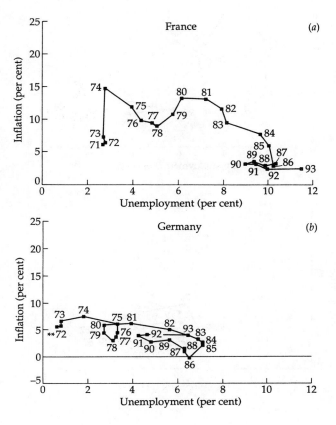

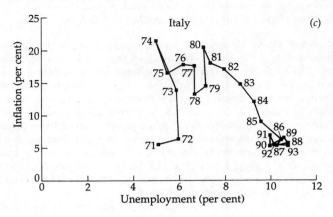

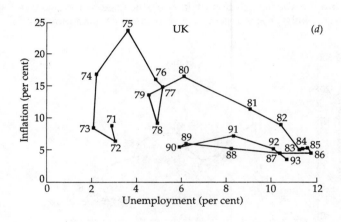

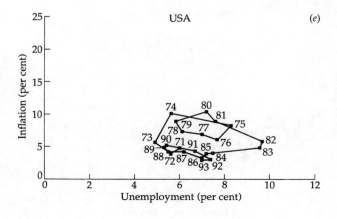

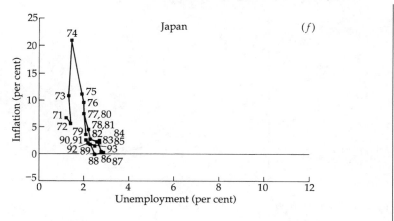

Fig. B3.1. Inflation and unemployment in major industrial countries, 1970–1993

3. Devaluation, time consistency, and credibility

The idea that when the government follows particular policies it plays a game with the private sector has conquered macroeconomic theory since the publication of the path-breaking articles of Kydland and Prescott (1977) and Barro and Gordon (1983).[12] This literature stresses that economic agents follow optimal strategies in response to the strategies of the authorities, and that these private sector responses have profound influences on the effectiveness of government policies. In particular, the reputation governments acquire in pursuing announced policies has a great impact on how these policies are going to affect the economy.

This literature also has important implications for our discussion of the costs of a monetary union. It leads to a fundamental criticism of the view that the exchange rate is a policy tool that governments have at their disposal to be used in a discretionary way. In order to understand this criticism it will be useful to present first the Barro-Gordon model for a closed economy, and then to apply it to an open economy, and to the choice of countries whether or not to join a monetary union.

3.1. The Barro-Gordon model: a geometric interpretation

Let us start from the standard Phillips curve which takes into account the role of inflationary expectations. We specify this Phillips curve as follows:

$$U = U_N + a(\dot{p}^e - \dot{p}) \tag{2.4}$$

---

[12] The Barro-Gordon model has been applied to open economies by Mélitz (1988) and Cohen and Wyplosz (1989) among others.

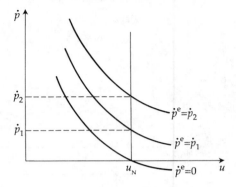

Fig. 2.8. The Phillips curve and natural unemployment

where U is the unemployment rate, U_N is the natural unemployment rate, $\dot{p}$ is the observed rate of inflation, and $\dot{p}^e$ is the expected rate of inflation.

Equation (2.4) expresses the idea that only unexpected inflation affects the unemployment rate. Thus, when the inflation rate $\dot{p}$ is higher than the expected rate of inflation, the unemployment rate declines below its natural level.

We will also use the rational expectations assumption. This implies that economic agents use all relevant information to forecast the rate of inflation, and that they cannot be systematically wrong in making these forecasts. Thus, *on average* $\dot{p} = \dot{p}^e$, so that *on average* $U = U_N$.

We represent the Phillips curve in Fig. 2.8.[13] The vertical line represents the 'long-term' vertical Phillips curve. It is the collection of all points for which $\dot{p} = \dot{p}^e$. This vertical line defines the natural rate of unemployment U_N which is also called the NAIRU (the non-accelerating-inflation rate of unemployment).

The second step in the analysis consists in introducing the preferences of the monetary authorities. The latter are assumed to care about both inflation and unemployment.

We represent these preferences in Fig. 2.9 in the form of a map of indifference curves of the authorities. We have drawn the indifference curves concave, expressing the idea that as the inflation rate declines, the authorities become less willing to let unemployment increase in order to reduce the inflation rate. Put differently, as the inflation rate declines the authorities tend to attach more weight to unemployment. Note also that the indifference curves closer to the origin represent a lower loss of welfare, and are thus preferred to those farther away from the origin.

---

[13] Note that we set the inflation rate on the vertical axis. This contrasts with the representation used in Sect. 2 of Ch. 1, where we had the rate of wage inflation on the vertical axis. We can, however, easily go from one representation to the other, considering that $\dot{w} = \dot{p} + \dot{q}$, where we assume that the rate of productivity growth, $\dot{q}$, is a constant.

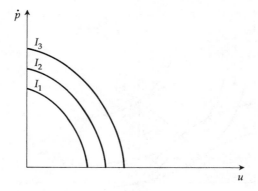

Fig. 2.9. The preferences of the authorities

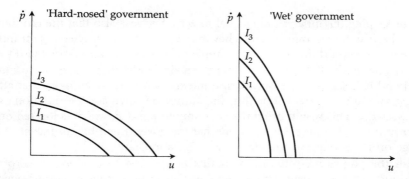

Fig. 2.10. The preferences of the authorities

The slope of these indifference curves expresses the relative importance the authorities attach to combating inflation or unemployment. In general, authorities who care much about unemployment ('wet' governments) have steep indifference curves, i.e. in order to reduce the rate of unemployment by one percentage point, they are willing to accept a lot of additional inflation.

On the other hand, 'hard-nosed' monetary authorities are willing to let the unemployment rate increase a lot in order to reduce the inflation rate by one percentage point. They have flat indifference curves. At the extreme, authorities who care only about inflation have horizontal indifference curves. We represent a few of these cases in Fig. 2.10.

We can now bring together the preferences of the authorities and the Phillips curves to determine the equilibrium of the model. We do this in Fig. 2.11.

In order to find out where the equilibrium will be located, assume for a moment that the government announces that it will follow a monetary policy

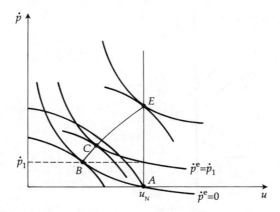

Fig. 2.11. The equilibrium inflation rate

rule of keeping the inflation rate equal to zero. Suppose also that the economic agents believe this announcement. They therefore set their expectations for inflation equal to zero. If the government implements this rule we move to point A.

It is now clear that the government can do better than point A. It could cheat and increase the rate of inflation unexpectedly. Thus, suppose that after having announced a zero inflation, the authorities increase the inflation rate unexpectedly. This would bring the economy to point B, which is located on a lower indifference curve. One can say that the government has an incentive to renege on its promise to maintain a zero inflation rate.

Will the government succumb to this temptation to engineer a surprise inflation? Not necessarily. The government also knows that economic agents are likely to react by increasing their expectations of inflation. Thus during the next period, the Phillips curve is likely to shift upwards if the government decides to increase the rate of inflation unexpectedly. The government should therefore evaluate the short-term gain from cheating against the future losses that result from the fact that the Phillips curve shifts upwards.

But suppose now that the government consists of short-sighted politicians who give a low weight to future losses, and that it decides to cheat. We then move to point B. This, however, will trigger a shift of the Phillips curve upwards. Given these new expectations, it will be optimal for the authorities to move to point C. This will go on until we reach point E. This point has the following characteristics. First, it is on the vertical Phillips curve, so that agents' expectations are realized. They have therefore no incentives any more to change their expectations further. Secondly, at E the authorities have no incentive any more to surprise economic agents with more inflation. A movement upwards along the Phillips curve going through E would lead to an indifferent curve located higher and therefore to a loss of welfare.

Point E can also be interpreted as the equilibrium that will be achieved in a rational expectations world when the authorities follow a *discretionary* policy, i.e. when they set the rate of inflation optimally each period given the prevailing expectations.

It is clear that this equilibrium is not very attractive. It is however the only equilibrium that can be sustained, given that the authorities are sufficiently short-sighted, and that the private sector knows this. The zero inflation rule (or any other constant inflation rule below the level achieved at *E*) has no credibility in a world of rational economic agents. The reason is that these economic agents realize that the authorities have an incentive to cheat. They will therefore adjust their expectations up to the point where the authorities have no incentive to cheat any more. This is achieved at point *E*. A zero inflation rule, although desirable, will not come about automatically.[14]

It should be stressed that this model is a static one. If the policy game is repeated many times, the government will have an incentive to acquire a reputation of low inflation. Such a reputation will make it possible to reach a lower inflation equilibrium. One way the static assumption can be rationalized is by considering that in many countries political institutions favour short-term objectives for politicians. For example, the next election is never far away, leading to uncertainty whether the present rulers will still be in place next period. Thus, what is implicitly assumed in this model is that the political decision process is inefficient, leading politicians to give a strong weight to the short-term gains of inflationary policies. The politicians as individuals are certainly as rational as private agents; the political decision process, however, may force them to give undue weight to the very short-term results of their policies.

Before analysing the question of how a monetary union might help the authorities to move to a more attractive equilibrium, it is helpful to study what factors determine the exact location of the 'discretionary' equilibrium (point *E*).

We distinguish two factors that affect the location of the discretionary equilibrium, and therefore also the equilibrium level of inflation.

(a) The preferences of the authorities. In Fig. 2.12 we present the cases of the 'wet' (steep indifference curves) and the 'hard-nosed' (flat indifference curves) governments. Assuming that the Phillips curves have the same slopes, Fig. 2.12 shows that in a country with a 'wet' government, the equilibrium inflation will be higher than in the country with a 'hard-nosed' government.

Note also that the only way a zero rate of inflation rule can be credible is when the authorities show no concern whatsoever for unemployment. In that case the indifference curves are horizontal. The authorities will choose the lowest possible horizontal indifference curve in each period. The inflation equilibrium will then be achieved at point *A*.[15]

---

[14] In the jargon of the economic literature it is said that the policy rule of zero inflation is 'time-inconsistent', i.e. the authorities face the problem each period that a better short-term outcome is possible. The zero inflation rule is incentive-incompatible.

[15] Rogoff (1985*b*) has suggested that the best thing that could happen to a country is that its monetary policy be run by an orthodox central banker.

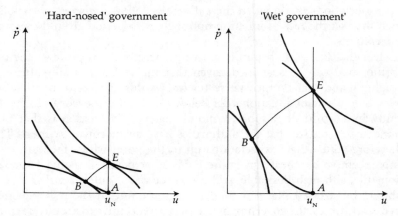

Fig. 2.12. Equilibrium with 'hard-nosed' and 'wet' governments

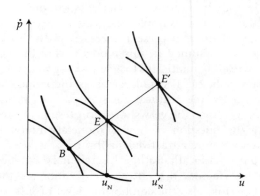

Fig. 2.13. Equilibrium and the level of natural unemployment

(b) The level of the natural rate of unemployment. Suppose the level of the natural unemployment rate increases. It can then easily be shown that if the preferences of the authorities remain unchanged, the new equilibrium inflation rate increases. This is made clear in Fig. 2.13 which shows the case of an increase of the NAIRU. Its effect is to shift the equilibrium point from *E* to *E'*.

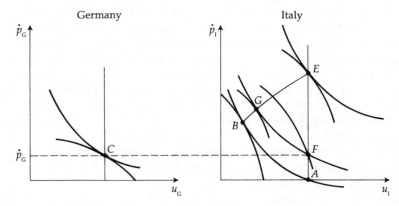

Fig. 2.14. Inflation equilibrium in a two-country model

3.2 The Barro-Gordon model in open economies

In the previous sections we showed how a government, which is known to care about inflation and unemployment, will not credibly be able to announce a zero inflation rate. It is therefore stuck in a suboptimal equilibrium with an inflation rate that is too high.

This analysis can be extended to open economies. Let us now assume that there are two countries. We call the first country Germany, and assume its government is 'hard-nosed'. The second country is called Italy, where the government is 'wet'. As in the model presented in Section 2 of Chaper 1, we use the purchasing-power parity condition, i.e.

$$\dot{e} = \dot{p}_{\mathrm{I}} - \dot{p}_{\mathrm{G}} \tag{2.5}$$

We show the inflation outcome in Fig. 2.14. Italy has a higher equilibrium rate of inflation than Germany. Its currency will therefore have to depreciate continuously. The problem of Italy is that it could achieve a much lower inflation equilibrium than point *E* if its government were able to convince its citizens that, once at point *A*, it would not try to reach point *B*.

This Barro-Gordon model for open economies allows us to add important insights into the discussion of the costs of a monetary union.

3.3. Credibility and the cost of a monetary union

Can Italy solve its problem by announcing that it will join a monetary union with Germany? In order to answer this question, suppose, first, that Italy announces that it will fix its exchange rate with the German mark. Given the purchasing-power parity, this fixes the Italian inflation rate at the German level. In Fig. 2.14 we show this by the horizontal line from C. Italy appears now to be able to enjoy a lower inflation rate. The potential welfare gains are large, because in the new equilibrium the economy is on a lower indifference curve.

The question, however, is whether this rule can be credible. We observe that once at the new equilibrium point F, the Italian authorities have an incentive to engineer a surprise devaluation of the lira. This surprise devaluation leads to a surprise increase in inflation and allows the economy to move towards point G. Over time, however, economic agents will adjust their expectations, so that the equilibrium inflation rate ends up to be the same as before the exchange rate was fixed. Thus, merely fixing the exchange rate does not solve the problem, because the fixed exchange rate rule is no more credible than a fixed inflation rate rule.[16]

There are, however, other arrangements that can potentially solve the high-inflation problem in Italy. Imagine that Italy decided to abolish its currency and to adopt the currency of Germany. If that arrangement could be made credible, i.e. if the Italian citizens were convinced that once this decision is taken and the mark becomes the national money, the Italian authorities would never rescind this decision, then Italy could achieve the same inflation equilibrium as Germany. In Fig. 2.14 the horizontal line connecting the German inflation equilibrium with Italy defines a credible equilibrium for Italy. The point F is now the new Italian inflation equilibrium. Since Italy has no independent monetary policy any more, its monetary authorities (with 'wet' preferences) have ceased to exist and therefore cannot devalue the lira. In the words of Giavazzi and Pagano (1987), Italy has borrowed credibility from Germany, because its government has its monetary hands firmly tied.[17]

This is certainly a very strong result. It leads to the conclusion that there is a large potential gain for Italy in joining a monetary union with Germany. In addition, there is no welfare loss for Germany. Thus, a monetary union only leads to gains. This analysis has become very popular especially in Latin countries where the distrust for one's own authorities runs very deeply.

There are two considerations, however, that tend to soften this conclusion. First, it should be clear from the previous analysis that only a full monetary union establishes the required credibility for Italy. That is, Italy must be willing to eliminate its national currency, very much like East Germany did on 1 July 1990 when it adopted the West German mark. Anything less than full monetary union will face a credibility problem. As was pointed out, when Italy fixes its exchange rate relative to the mark and keeps its own currency, as it does in the EMS, the credibility of this fixed exchange rate arrangement will be in doubt. We return to this question when we analyse the workings of the European Monetary System.

[16] We will come back to this issue when we discuss the workings of the EMS. Some economists have argued that fixing the exchange rate can be a rule that inherently has more credibility than announcing a constant inflation rate rule.

[17] See also Giavazzi and Giovannini (1989).

Secondly, and more importantly for our present purpose, we have assumed that the central bank of the monetary union is the German central bank. In this arrangement, Italy profits from the reputation of the German central bank to achieve lower inflation. Suppose, however, that the new central bank is a new institution, where both the German and the Italian authorities are represented equally. Would that new central bank have the same reputation as the old German central bank? This is far from clear. If the union central bank is perceived to be less 'hard-nosed' than the German central bank prior to setting up the union, the new inflation equilibrium of the union will be higher than the one which prevailed in Germany before the union. Italy may still gain from such an arrangement. Germany, however, would lose, and would not be very enthusiastic to form such a union. We return to these issues in Chapter 8 where we discuss problems of setting up a European central bank, and of devising institutions that will enhance the low-inflation reputation of the European central bank.

We conclude from the preceding analysis that problems of credibility are important in evaluating the costs of a monetary union. First, the option to devalue the currency is a two-edged sword for the national authorities. The knowledge that it may be used in the future greatly complicates macroeconomic policies. Secondly, the time-consistency literature also teaches us some important lessons concerning the costs of a monetary union: a devaluation cannot be used to correct every disturbance that occurs in an economy. A devaluation is not, as it is in the analysis of Mundell, a flexible instrument that can be used frequently. When used once, it affects its use in the future, because it engenders strong expectational effects. It is a dangerous instrument that can hurt those who use it. Each time the policy-makers use this instrument, they will have to evaluate the advantages obtained today against the cost, i.e. that it will be more difficult to use this instrument effectively in the future.

This has led some economists to conclude that the exchange rate instrument should not be used at all, and that countries would even gain from irrevocably relinquishing its use. This conclusion goes too far. There were many cases, observed in Europe during the 1980s, in which devaluations were used very successfully (see the previous section). The ingredients of this success have typically been that the devaluation was coupled with other drastic policy changes (sometimes with a change of government, e.g. Belgium in 1982 and Denmark in the same year). As a result, the devaluation was perceived as a unique and an extraordinary change in policies that could not easily be repeated in the future. Under those conditions the negative reputation effects could be kept under control. Some countries, in particular Denmark in 1982, even seem to have improved their reputation quickly after the devaluation. Relinquishing the possibility of using this instrument for the indefinite future does imply a cost for a nation.

4. The cost of monetary union and the openness of countries

In this chapter we have developed several ideas that bear on the question of how the openness of a country affects the cost of the monetary union. Here we concentrate on two of these which, as will be seen, have opposite effects. First, there is the relation between the degree of openness (the degree to which a country is integrated with the rest of the world) and the occurrence of asymmetric shocks. We have presented two views: the European Commission view, which sees this as a negative relationship, and the Krugman view, which sees this as a positive one. According to the first view, we can conclude that more openness reduces the cost of a monetary union, as it reduces the probability that asymmetric shocks occur. On the second view, however, this conclusion is reversed: the costs of a monetary union increase with the degree of openness of countries.

The second idea which matters in our analysis of how openness affects the cost of monetary union has to do with the effectiveness of the exchange rate in dealing with asymmetric shocks. Let us return to Fig. 2.5, where we analysed the effects of a devaluation. We now consider two countries, one relatively open, the other relatively closed. We represent these two countries in Fig. 2.15. Both the demand and the supply effects of a devaluation differ between the two countries. As far as the demand-side effects are concerned, the same devaluation has a stronger effect in the relatively open economy than in the relatively closed one. To understand this, consider two extreme cases. Suppose the relatively open economy exports 99% of its GDP, whereas the relatively closed one only exports 1% of its GDP. The same devaluation, say 10%, is bound to raise aggregate demand more in the former country than in the latter. In Fig. 2.15 we show this difference by the fact that the demand curve in the relatively open country shifts further outward than in the relatively closed country.

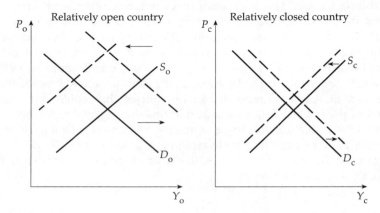

Fig. 2.15. Effectiveness of devaluation as a function of openness

The two countries also differ with respect to the supply-side effects of the devaluation. One can expect that in the relatively open economy, the upward shift of the supply curve following the devaluation is more pronounced than in the relatively closed economy. This has to do with the fact that the more open economy imports more (as a per cent of total consumption) so that the CPI increases more, leading to a stronger wage–price spiral than in the relatively closed economy.

We now arrive at the following conclusion. The combined demand and supply effects of a devaluation in the two countries are such that we cannot say a priori in which country the devaluation is most effective in stimulating output. What we can conclude, however, is that the devaluation will be felt more strongly on the aggregate price level in the more open economy than in the relatively closed economy. This means that the systematic use of the exchange rate instrument will lead to more price variability in the more open economy than in the relatively closed one. To the extent that price variability involves costs, the systematic use of the exchange rate instrument in the more open economy will be more costly. This point was first recognized by McKinnon (1963) in his important contribution to the theory of optimum currency areas.

From this discussion of the effectiveness of a devaluation in open and (relatively) closed economies one can conclude that for the same effect on output, the devaluation is likely to be more costly in the open economy because of the higher price variability involved. Thus the loss of the exchange rate instrument is likely to be less costly for the relatively open economy than for the relatively closed one.

Combining the analysis of the effectiveness of a devaluation with the analysis of the relationship between openness and asymmetric shocks, one can derive the conclusion that the cost of monetary union most likely declines with the degree of openness of a country. We show this in Fig. 2.16, which is borrowed

Fig. 2.16. The cost of a monetary union and the openness of a country

from Krugman (1990). On the vertical axis the cost of a monetary union is set out (i.e. the cost of relinquishing the exchange rate instrument). This cost is expressed as a per cent of GDP. On the horizontal axis the openness of the country relative to the countries with whom it wants to form a monetary union is set out. This openness is represented by the trade share in the GDP of the country considered here. We see that as the openness increases the cost of joining a monetary union declines.

This conclusion, however, does not hold in general. We cannot exclude the possibility that the relationship is positively sloped. This situation may occur when the probability of asymmetric shocks increases significantly with the degree of openness of a country (the Krugman scenario). The presumption, however, is that this case is unlikely to happen. First, it is more likely that with trade integration, asymmetric shocks become less likely (see our analysis of the European Commission view). Second, even if trade integration leads to more asymmetric shocks, the cost of using the exchange rate instrument is likely to offset the former effect in more highly integrated economies. We will therefore maintain our presumption that the cost of the monetary union declines with openness.

5. Conclusion

The criticism against the traditional theory of optimal currency areas, as developed by Mundell and McKinnon, has enabled us to add important nuances to this theory. In particular, it has changed our view about the costs of a monetary union. The traditional theory of optimal currency areas tends to be rather pessimistic about the possibility for countries to join a monetary union at a low cost. The criticism we have discussed in this chapter is much less pessimistic about this, i.e. the costs of forming a monetary union appear to be less forbidding.[18] Despite this criticism the hard core of the optimum currency analysis still stands. This can be put as follows:

(1) There are important differences between countries that are not going to disappear in a monetary union. This raises the question whether countries should gladly relinquish control over their ability to follow an independent monetary policy, including setting (and changing) the price of their currency.

(2) Despite the fact that exchange rate changes usually have no permanent effects on real variables, such as output and employment, variations in the exchange rate remain a powerful instrument to help countries to eliminate important macroeconomic disequilibria, and to make the adjustment process less costly in terms of lost output and employment. The examples of France, Belgium, and Denmark that devalued their currencies during the 1980s (and coupled this devaluation with the right domestic policies) illustrate this point.

---

[18] See Tavlas (1993) for some further thoughts about this.

(3) The argument that exchange rate changes are dangerous instruments in the hands of politicians is important. The experience of many countries illustrates that when devaluations are used systematically, they will lead to more inflation without gains in terms of output and employment. In addition, they easily lead to macroeconomic instability as economic agents continuously tend to expect future devaluations. Contrary to the old view, devaluations are not instruments that policy-makers can use flexibly and costlessly.

This argument should, however, not be pushed too far. The fact that such an instrument can be misused is not sufficient reason to throw it away, when it can also be put to good use, when countries face extraordinary circumstances. The abandonment of the exchange rate instrument to deal with possible extraordinary disturbances should certainly be considered to be a cost of monetary union.

3

THE BENEFITS OF A COMMON CURRENCY

Introduction

Whereas the costs of a common currency have much to do with the *macro-economic* management of the economy, the benfits are mostly situated at the *microeconomic* level. Eliminating national currencies and moving to a common currency can be expected to lead to gains in economic efficiency. These gains in efficiency have two different origins. One is the elimination of transaction costs associated with the exchanging of national moneys. The other is the elimination of risk coming from the uncertain future movements of the exchange rates. In this chapter we analyse these two sources of benefits of a monetary union.

1. Direct gains from the elimination of transaction costs

Eliminating the costs of exchanging one currency into the other is certainly the most visible (and most easily quantifiable) gain from a monetary union. We have all experienced these costs whenever we exchanged one currency for another. These costs disappear when countries move to a common European currency.

How large are these gains from the elimination of transaction costs? The EC Commission has estimated these gains, and arrives at a number between 13 and 20 billion ECUs per year.[1] This represents one-quarter to one-half of one per cent of the Community GDP. This may seem peanuts. It is, however, a gain that has to be added to the other gains from a single market.

It should be noted here that these gains that accrue to the general public have a counterpart somewhere. They are mostly to be found in the banking sector. Surveys in different countries indicate that about 5% of the banks'

[1] See EC Commission (1990).

revenues are the commissions paid to the banks in exchange of national currencies. This source of revenue for the banks will disappear with a monetary union.

The preceding should not give the impression that the gain for the public is offset by the loss of the banks. The transaction costs involved in exchanging money are a *deadweight* loss. They are like a tax paid by the consumer in exchange for which he gets nothing. Banks, however, will have a problem of transition: they will have to look for other profitable activities. When this has been done, society will have gained. The banks' employees, previously engaged in exchanging money, will now be free to perform more useful tasks for society.

Another point to be made here is that the gains from the elimination of transaction costs can only be reaped when national moneys are replaced by a common currency. These gains are unlikely to occur if in a future monetary union national moneys would remain in place, albeit with irrevocably fixed exchange rates. As long as national currencies remain in existence, even if the exchange rate is 'irrevocably' fixed, doubts will continue to exist as to this fixity. This will stimulate residents of each country to use their home currency in preference to foreign ones. In other words, currencies will not be perfect substitutes. There will continue to be a need to convert one currency into another. Those that provide this service will charge a price. Transaction costs will not be eliminated.

2. Indirect gains from eliminating transaction costs

The elimination of transaction costs will also have an indirect (albeit less easily quantifiable) gain. It will reduce the scope for price discrimination between national markets.

There is a lot of evidence that price discrimination is still practised widely in Europe. In Table 3.1 we illustrate this phenomenon in the automobile market. It can be seen that the same automobiles are priced very differently (net of taxes) in the European Union.

Such price discrimination is only possible because national markets are still segmented. That is, there are relatively large transaction costs for the consumer who would buy a car in another country. If these transaction costs did not exist, consumers would not hesitate to purchase these goods in the countries where they are cheap. Of course, there are many sources of transaction costs (e.g. administrative regulations, differences in taxation), and eliminating the cost of buying and selling foreign currencies may not even be the most important one. However, together with the other measures to create a single market, they would make price discrimination much more difficult. This would be a benefit for the European consumer.

The importance of the existence of national currencies in segmenting markets should not be underestimated. In a recent study Charles Engel and

Table 3.1. Average price differentials (net of taxes) for the same automobile in Europe, 1993 and 1995 (cheapest country = 100)

	1993	1995
Belgium	116	122
France	121	121
Germany	124	128
Ireland	115	112
Italy	100	100
Netherlands	115	121
Portugal	108	108
Spain	108	105
United Kingdom	120	120

Source: European Commission (various years).

Richard Rogers[2] analysed the factors that influence the price differentials of the same goods in different locations. They did this by studying the price differentials of the same pairs of goods in different North American cities (in the US and Canada). What they found is quite revealing. First, distance matters. Price differentials between Los Angeles and New York are larger than between Los Angeles and San Francisco (this is not really surprising). Second, and more importantly, crossing a border (in this case the US–Canadian border) is equivalent to travelling 2,500 miles within the same country. In other words, price differentials between Detroit and Windsor (which is just across the border) are of the same order of magnitude as the price differentials between New York and Los Angeles. Thus, borders are quite powerful in segmenting markets and in introducing large variations in the movements of prices. Of course, crossing borders not only involves exchanging moneys. As mentioned earlier, borders create other impediments to trade. Nevertheless, the fact that at borders moneys have to be exchanged is a significant factor in explaining why markets remain segmented.

The previous discussion also makes clear that a monetary union will have a great potential further to integrate markets in the European Union, in the same way as having the same currency, the dollar, has been of great significance for the United States in creating a truly single market in that country.

3. Welfare gains from less uncertainty

The uncertainty about future exchange rate changes introduces uncertainty about future revenues of firms. It is generally accepted that this leads to a loss of welfare in a world populated by risk-averse individuals. These will, generally speaking, prefer a future return that is more certain than one that is less so, at

---

[2] See Engel and Rogers (1995).

Price certainty

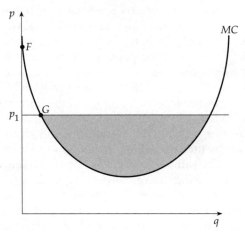

Price uncertainty

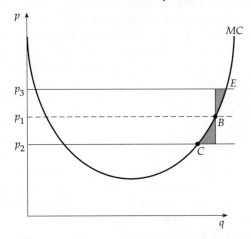

Fig. 3.1. Profits of the firm under price certainty and uncertainty

least if the expected value of these returns is the same. Put differently, they will only want to take the more risky return if they are promised that it will be higher than the less risky. Eliminating the exchange risk reduces a source of uncertainty and should therefore increase welfare.

There is one important feature of the theory of the firm that may invalidate that conclusion. Take a profit-maximizing firm which is a price-taker in the output market. We represent its marginal cost curve and the price of its output in Fig. 3.1.

Suppose there are two regimes. In the first regime (presented in the left-hand panel) the price is constant and perfectly predictable by the firm. In the second regime (right-hand panel) the price fluctuates randomly. We assume here that the price fluctuates symmetrically between p_2 and p_3.

In the first regime of certainty the profit of the firm in each period is given by the shaded area minus the area FGp_1. In the second uncertain regime the profit will fluctuate depending on whether the price p_2 or p_3 prevails. We can now see that the profit will be larger on average in the uncertain regime than in the certain regime. When the price is low the profit is lower than in the certainty case by the area p_1BCp_2. When the price is high, the profit is higher than the certainty case by the area p_3EBp_1. It can now easily be seen that P_3EBp_1 is larger than p_1BCp_2. The difference is given by two darkened triangles.

This result has the following interpretation. When the price is high the firm increases output so as to profit from the higher revenue per unit of output. Thus, it gains a higher profit for each unit of output it would have produced anyway, and *in addition* it expands its output. The latter effect is measured by the upper darkened triangle. When the price is low, however, the firm will do the opposite, it will reduce output. In so doing it limits the reduction in its total profit. This effect is represented by the lower darkened triangle.

There are many complications that can be added to this theory. For example, one can introduce the assumption of imperfect competition, or the assumption that there are adjustment costs. In general, the conclusions may not be as sharp as in the simple case presented here. Nevertheless, in these more complicated models, it is generally the case that price uncertainty may increase the average profits of the firm.

If one wants to make welfare comparisons between a regime of price certainty and one of price uncertainty, the positive effect of price uncertainty on the average profits should be compared to the greater uncertainty about these profits. The higher average profit increases the utility of the firm, whereas the greater uncertainty about these profits reduces the utility of the (risk-averse) firm. It is, therefore, unclear whether welfare will decline when exchange rate uncertainty increases, or, conversely that we can say with great confidence that the welfare of firms will increase when national currencies are eliminated and a common currency is introduced.

Another way to put the preceding analysis is to recognize that changes in the exchange rate do not only represent a risk, they also create opportunities to make profits. When the exchange rate becomes more variable the probability of making very large profits increases. In a certain sense, exporting can be seen as an *option*. When the exchange rate becomes very favourable the firm exercises its option to export. With an unfavourable exchange rate the firm does not exercise this option. It is well known from option theory that the value of the option increases when the variability of the underlying asset increases. Thus, the firm that has the option to export is better off when the exchange rate becomes more variable.

The same argument can be developed for the *consumer*. In Fig. 3.2 we

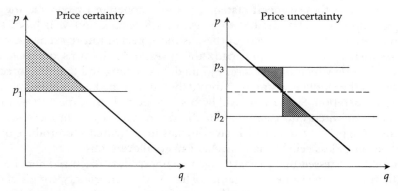

Fig. 3.2. Average consumer surplus and price volatility

present the demand function of a representative consumer. Suppose again that there are two regimes of price volatility. In the first regime the price is constant and perfectly predictable. In the second regime the price fluctuates randomly between p_2 and p_3. We find that in the second regime of price uncertainty the consumer surplus is higher on average than in the first regime of price certainty. The reason is the same as in the case of the firm. When the price is low, the consumer increases his demand to profit from the low price. When the price is high he does the opposite, and thereby limits the negative effect the price increase has on his welfare. Thus, on average, the consumer gains when prices fluctuate.

This positive effect of price uncertainty on the average consumer surplus has to be compared to the increased risk. To the extent that consumers are risk-averse they will give a lower utility value to the higher (but uncertain) consumer surplus than to the lower consumer surplus obtained in a regime of certainty. Our conclusion from the static theory of the consumer is that we do not know whether consumers gain from lower price variability. We also conclude that if gains from a common currency and the ensuing reduction in risk are to be expected, they should probably be found elsewhere than in the static welfare gains that we have analysed in this section.

4. Exchange rate uncertainty and the price mechanism

There is another area where more substantial gains from a reduction of the exchange rate risk can be expected. Exchange rate uncertainty introduces uncertainty about the future prices of goods and services. Economic agents base their decisions concerning production, investment, and consumption on the information that the price system provides for them. If these prices become more uncertain the quality of these decisions will decline.

We can make these general statements more concrete by considering an example. Suppose a firm decides to invest in a foreign country. It bases this decision on many variables. One of these is the expected future exchange rate. Suppose then that after having made the investment, it turns out that the exchange rate on which the decision was made was wrong and that this forecast error makes the whole investment unprofitable so that the firm decides to close its foreign operation. Such errors will be costly. One can also expect them to be more frequent when uncertainty about the future exchange rate increases. In this sense, the price system, which gives signals to individuals to produce or to invest, becomes less reliable as a mechanism to allocate resources.

It should be stressed that the exchange rate uncertainty discussed here has to do with *real* exchange rate uncertainty. That is, the uncertainty comes about because the exchange rate changes do not reflect price changes. A well-known example is the dollar appreciation during 1980–5, which was largely unpredicted and which went much farther than the inflation differential between the US and the other industrial countries. In other words, the dollar deviated substantially from its purchasing power parity. This 'misalignment' led to large and unpredicted changes in the profitability of many American industrial firms which had to compete in world markets. It also led to declines in output and firm closures. A few years later the dollar depreciated substantially and more than corrected the real appreciation of the first half of the 1980s. These large real exchange rate movements led to large adjustment costs for the American economy. (For a well-known analysis of the effects of misalignment see Williamson (1983)).

A decline in real exchange rate uncertainty, due for example to the introduction of a common currency, can reduce these adjustment costs. As a result, the price system becomes a better guide to make the right economic decisions. These efficiency gains are difficult to quantify. They are no less important for this. Their importance becomes all the more visible when we look at what happens in countries that experience hyperinflation. We observe in these countries that the wrong production and investment decisions are made on a massive scale. Quite often we observe that output and investment booms in countries with hyperinflation. However, these increases in output and investment frequently occur in the wrong sectors or product lines. After a while these productions and investments have to be abandoned. Massive amounts of resources are wasted in the process.

There is a second reason why greater price and exchange rate uncertainty may reduce the quality of the information provided by the price mechanism. An increase in risk, due to price uncertainty, will in general increase the real interest rate. This follows from the fact that when the expected return on investment projects becomes more uncertain, risk-averse investors will require a higher risk premium to compensate them for the increased riskiness of the projects. In addition, in a riskier economic environment, economic agents will increase the discount rate at which they discount future returns. Thus, exchange rate uncertainty which leads to this kind of increased systemic risk also increases the real interest rate. Higher interest rates, however, lead to increased problems in selecting investment projects in an efficient way. These problems have to do with *moral hazard* and *adverse selection*.

The *moral hazard problem* arises because an increase in the interest rate changes the incentives of the borrower. The latter will find it more advantageous to increase the riskiness of his investment projects. This follows from an asymmetry of expected profits and losses. If the investment project is successful the extra profits go to the borrower. If the investment project turns out badly and if the borrower goes bankrupt, his loss is limited to his equity share in the project. With a higher interest rate this moral hazard problem becomes more intense. This asymmetry gives the borrower the incentive to select more risky projects. Thus, on average, investment projects will become riskier when the real interest rate increases. Lenders, however, will try to defend themselves by asking for an additional risk premium, which in turn intensifies the problem. In general, the moral hazard problem may lead the lender to apply credit ceilings as a way to reduce his risk.[3]

The *adverse selection problem* leads to a similar result. When the interest rate increases, the suppliers of low-risk investment projects will tend to drop out of the credit market. They will find it less attractive to borrow at the higher interest rate for projects that do not represent a high risk. Thus, on average, the riskiness of investment projects will increase when the interest rate increases.

Both phenomena, moral hazard and adverse selection, lead to the selection of more risky investment projects. Thus, the systemic risk increases. Eliminating that risk by moving towards a common currency reduces the amount of risky projects that are selected by the market.

We can conclude this section by noting that the movement towards a common currency will eliminate the exchange risk, and thereby will lead to a more efficient working of the price mechanism. Although this effect cannot easily be measured, it is likely to be an important benefit of the introduction of one currency in Europe.

It should be noted here that not all economists will share this conclusion. Some have argued that the elimination of the exchange risk can only be obtained by introducing more risk elsewhere in the economic system. As a result, we are not sure whether the systemic risk is reduced by eliminating just one source of risk. We return to this issue in Box 4 where we evaluate this argument.

5. Exchange rate uncertainty and economic growth

The argument that the elimination of the exchange risk will lead to an increase in economic growth can be made using the neo-classical growth model, and its recent extension to situations of dynamic economies of scale. This analysis

---

[3] See the classic article of Stiglitz and Weiss (1981).

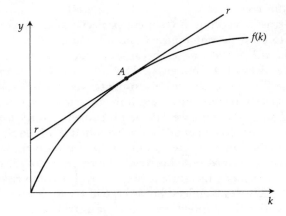

Fig. 3.3. The neoclassical growth model

features prominently in the EC report 'One Market, One Money' (1990) which in turn was very much influenced by Baldwin (1989).

The neoclassical growth model is represented in Fig. 3.3. The horizontal axis shows the capital stock per worker, the vertical axis the output per worker. The line $f(k)$ is the production function which has the usual convex shape, implying diminishing marginal productivities. The equilibrium in this model is obtained where the marginal productivity of capital is equal to the interest rate consumers use to discount future consumption. This is represented in Fig. 3.3 by the point A, where the line rr (whose slope is equal to the discount rate) is tangent to the production function $f(k)$. In this model, growth can only occur if the population grows or if there is an exogenous rate of technological change. (Note also that in this neoclassical model the savings ratio does not influence the equilibrium growth rate.)

We can now use this model as a starting-point to evaluate the growth effects of a monetry union. Assume that the elimination of the exchange risk reduces the systemic risk so that the real interest rate declines. We represent this effect in Fig. 3.4. The reduction of the risk-adjusted rate of discount makes the rr-line flatter. As a result, the equilibrium moves from A to B. There will be an accumulation of capital and an increase in the growth rate while the economy moves from A to B. In the new equilibrium, output per worker and the capital stock he has at his disposal will have increased. Note, however, that the growth rate of output then returns to its initial level, which is determined by the exogenous rate of technological change and the rate of growth of the population. Thus, in this neoclassical growth model the reduction of the interest rate due to the monetary union *temporarily* increases the rate of growth of output. In the new equilibrium the output *level* per worker will have increased. (Note also that the productivity of capital has declined.)

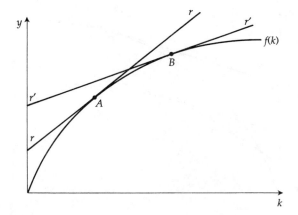

Fig. 3.4. The effect of lower risk in the neoclassical growth model

This model has been extended by introducing dynamic economies of scale.[4] Suppose the productivity of capital increases when the capital stock increases. This may arise because with a higher capital stock and output per worker there are learning effects and additional knowledge is accumulated. This additional knowledge then increases the labour productivity in the next period. There may also be a public goods aspect to knowledge. Thus, once a new machine is in place the knowledge it embodies is freely available to the worker who uses it. All these effects produce increases in the productivity of labour over time when capital accumulates.

One of the interesting characteristics of these new growth models is that the growth path becomes endogenous, and is sensitive to the initial conditions. Thus, an economy that starts with a higher capital stock per worker can move on a permanently higher growth path.

A lowering of the interest rate can likewise put the economy on a permanently higher growth path. We represent this case in Fig. 3.5. As a result of the lower interest rate the economy accumulates more capital. Contrary to the static case of Fig. 3.4, however, this raises the productivity of the capital stock per worker. This is shown by the upward movement of the $f(k)$ line. The economy will be on a higher growth path.

The previous analysis sounds very promising for the growth effects of a monetary union.[5] It is, however, most probably a little too optimistic, for it ignores the point that was made in Section 3 above, that is, a reduction of

---

[4] See Romer (1986). What is new here is the formalization of old ideas. Many economists in the past have stressed that the growth phenomenon is based on dynamic economies of scale and learning effects.

[5] This analysis was also implicit in the hope of the founding fathers of the EMS that the greater exchange rate stability provided by the system would stimulate the growth of investment, output, and trade in Europe.

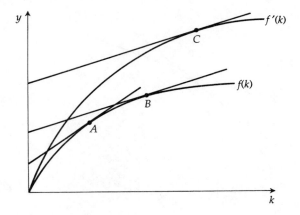

Fig. 3.5. Endogenous growth in the 'new' growth model

exchange rate variability also reduces the expected value of future profits of firms. Thus, a lower risk due to less exchange rate variability has a double effect. It reduces the real interest rate (which in the previous analysis generated the growth effect) *and* it reduces the expected return of investment. As a result, this reduction in risk has an ambiguous effect on investment activity, and thus also on the growth of output.

This point is not generally recognized by the public. Quite often it is stated that a reduction of the risk will boost investment activity. Economic theory does not allow us to draw this conclusion. The question whether a reduction of the exchange risk increases investment, therefore, is an empirical one. What does the empirical evidence tell us?

There has been a large amount of empirical analysis of the relation between exchange rate uncertainty and international trade and investment. (For a survey, see IMF (1984).) On the whole it is fair to say that very little relation has been found. In other words, the increased variability of the exchange rates, and in particular the large and unpredictable variability of the real exchange rates, does not seem to have had very significant effects on international trade and investment. This implies that the link between exchange rate uncertainty and economic growth is empirically also a very weak one.

We illustrate this lack of empirical relationship between exchange rate variability and the growth of output and investment in Table 3.2. We show the growth rates of GDP and of investment in industrial countries during the period when the EMS (with narrow bands) existed. We classify these countries into two groups, those that have experienced relatively stable (nominal and real) exchange rates (mainly the EMS countries) and those that have seen their (nominal and real) exchange rates fluctuate a lot. In general, the latter countries have experienced exchange variability (both nominal and real) that is three to five times as large as the former.

Table 3.2. Growth of GDP and investment, 1981–1993

	Investment	GDP
EMS countries	*0.94*	*2.10*
Austria	2.15	2.07
Belgium	1.54	1.65
Denmark	0.52	1.80
France	1.11	1.87
Germany	0.92	2.12
Ireland	−0.43	3.48
Italy	0.32	1.75
Netherlands	1.36	2.04
Non-EMS countries	*1.72*	*2.14*
Finland	−1.63	1.49
Greece	−0.07	1.44
Japan	4.35	3.57
Portugal	2.52	2.45
Spain	3.17	2.45
Sweden	−0.04	1.19
United Kingdom	2.52	2.02
USA	2.92	2.50

Source: EC Commission, *European Economy*, 62 (1996).
Note: Spain, Portugal, and the UK belonged to the EMS during a short
period from 1990 on.

The figures of Table 3.2 are striking. The greater exchange rate stability that the EMS countries have experienced during the 1980s does not seem to have provided a great boost to the growth rates of output and investment.[6] As a matter of fact, the growth rates of output and investment have on average been somewhat lower in the EMS countries than in the non-EMS countries that experienced relatively large movements in their exchange rates.

Another way to illustrate this lack of a robust relationship between economic growth and exchange rate risk is to look at the growth rate of countries as a function of their size. Large countries have a large monetary zone within which there is no exchange rate uncertainty. Firms in small countries typically face much more exchange rate uncertainty because they sell a larger proportion of their final output to countries in different monetary areas. Thus, a large part of these sales face exchange rate uncertainty. Therefore one would expect that if a reduction of exchange rate uncertainty stimulates economic growth, larger countries will on average experience a higher growth rate of output than small countries. We show some evidence in Fig. 3.6. On the vertical axis we present the growth rates during 1972–93, on the horizontal axis the size of these countries as measured by their GDP (in 1993). It is immediately clear that there is no relationship between the size of countries and their growth rates.

[6] For more evidence on the growth effects of the EMS see De Grauwe (1987). Note that it is not implied here that the greater exchange rate stability observed in the EMS has not been beneficial for the EMS countries. What is implied is that this greater exchange rate stability does not seem to have had much beneficial effect on the growth rates of output and investment.

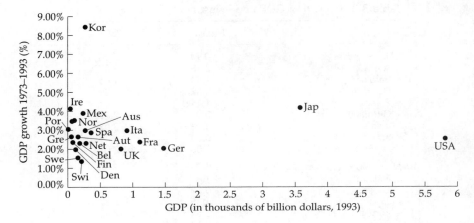

Fig. 3.6. Growth rates of GDP (1972–1993) and level of GDP (1993)

Sources: International Monetary Fund, *IFS*; OECD, *Main Economic Indicators*.

The previous evidence is of course only meant to be suggestive. Most of the many econometric studies that have been performed recently tend to confirm that the degree of exchange rate variability has a very weak impact on the growth rates of investment, trade, and output. Thus, the weakness of the theoretical argument for such a relationship is confirmed by a weakness of the empirical relationship.

There are, however, two other possible explanations for our failure to find a significant empirical relationship between exchange rate uncertainty and economic growth.

A first alternative explanation is that when we compare the experience of the EMS countries with the other countries, we fail to take into account the fact that the exchange rate uncertainty within the EMS, although reduced, has not been eliminated. It may be that the movement to full monetary union with a common currency is the step we need to eliminate the exchange rate uncertainty, and to stimulate economic growth. (Note, however, that this interpretation of the empirical results is less convincing when we look at the evidence concerning the size of countries.)

A second more promising explanation is that the reduction in exchange rate uncertainty may not necessarily reduce the systemic risk. Less exchange rate uncertainty may be compensated by greater uncertainty elsewhere, e.g. output and employment uncertainty. As a result, firms that face a greater monetary zone may not on average operate in a less risky environment. There is a whole theoretical literature, starting with William Poole (1970), that has analysed this problem. We have presented the main results in Box 4.

6. Benefits of a monetary union and the openness of countries

As in the chapter on the costs of a monetary union, we can also derive a relationship between the *benefits* of a monetary union and the openness of a country. The welfare gains of a monetary union that we have identified in this chapter are likely to increase with the degree of openness of an economy. For example, the elimination of transaction costs will weigh more heavily in countries where firms and consumers buy and sell a large fraction of goods and services in foreign countries. Similarly, the consumers and the firms in these countries are more subject to decision errors because they face large foreign markets with different currencies. Eliminating these risks will lead to a larger welfare gain (per capita) in small and open economies than in large and relatively closed countries.

We can represent this relationship between the benefits of a monetary union and the openness of the countries that are candidates for a union graphically. This is done in Fig. 3.7. On the horizontal axis we show the openness of the country relative to its potential partners of the monetary union (measured by the share of their bilateral trade in the GDP of the country considered). On the vertical axis we represent the benefits (as a percentage of GDP). With an increasing openness towards the other partners in the union, the gains from a monetary union (per unit of output) increase.

Fig. 3.7. Benefits of a monetary union and openness of the country

Box 4. Fixing exchange rates and systemic risk

In a path-breaking article William Poole (1970) showed that fixing interest rates does not necessarily reduce the volatility of output compared to fixing the money stock. The argument Poole developed can easily be extended to the choice between fixed and flexible exchange rates.

Suppose we can represent the economy by the standard *IS–LM* model. In an open economy with perfect capital mobility the domestic interest rate is equal to the foreign interest rate plus the expected future depreciation of the domestic currency (open interest parity).

We consider first *random shocks occurring in the goods market* (business cycle shocks for example). We present this by shifts in the *IS* curve. The latter now moves unpredictably between IS_U and IS_L.

Let us assume first that the authorities fix the exchange rate, and that this fixed exchange rate is credible, so that the domestic interest rate must be equal to the foreign interest rate. We represent the model graphically in Fig. B4.1.

The fixity of the exchange rate here implies that the domestic interest rate is constant (assuming no changes in the foreign interest rate). Thus, output will fluctuate between Y_L and Y_U. (Note that as the *IS* curve moves to, say, IS_U the *LM* curve will also automatically be displaced to the right, so that it intersects IS_U at the point *F*. This movement of the *LM* curve comes from the fact that the upward movement of the *IS* curve tends to increase the domestic interest rate: with perfect capital mobility this leads to capital inflows, which increase the domestic money stock).

What happens if the authorities allow the exchange rate to be flexible, and instead fix the domestic money stock? In this case the *LM* curve remains fixed. The same shocks in the *IS* curve now have no effect on the output level. The reason is the following. Suppose the *IS* curve shifts upwards (say because of a domestic boom). This tends to increase the domestic interest rate. Since the exchange rate is flexible, there can be no increase in the money stock from net capital inflows. Instead, the increase in the domestic interest rate leads to an

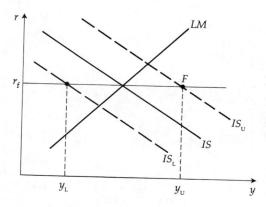

Fig. B4.1. Shocks in the *IS* curve

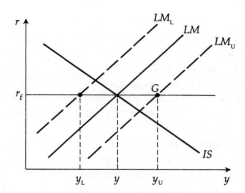

Fig. B4.2. Shocks in the *LM* curve

appreciation of the currency. This appreciation, in turn, tends to shift the *IS* curve back to the left. This will continue until the domestic interest rate returns to its initial level, which is only possible when the *IS* curve returns to its initial position.

We conclude from this case that fixing the exchange rate has led to more variability in the output market (and therefore also in the labour market) compared to letting the exchange rate vary. Fixing the exchange rate does not reduce systemic risk, because it leads to more uncertainty elsewhere in the system.

This result, however, very much depends on the nature of the random shocks, that were assumed to come from the goods markets. Things are quite different if the random shocks originate from the money market.

Suppose that we have *random disturbances in the demand for money* (disturbances in velocity). We represent these by movements in the *LM* curve between the limits *LM*$_L$ and *LM*$_U$ in Fig. B4.2. Let us again consider the case where the authorities fix the exchange rate. As before, this means that the domestic interest rate is fixed (assuming no shocks in the foreign interest rate). It can now immediately be established that there will be no change in output. The reason is the following. Suppose that the demand for money has declined leading to a rightward shift of the *LM* curve. This tends to reduce the interest rate. Such a reduction, however, is prevented by a capital outflow. At the same time the money stock declines. The *LM* curve must return to its initial level. Thus, the goods market is completely insulated from money market disturbances when the authorities fix the exchange rate.

If the authorities allow the exchange rate to float, this will not be the case any more. Output will now fluctuate between the levels Y_L and Y_U. The intuition is that if the *LM* curve shifts to the right, the ensuing decline in the interest rate leads to a depreciation of the currency, whereas the domestic money supply remains unchanged. The decline in the interest rate and the depreciation tend to stimulate aggregate demand. This shifts the *IS* curve upwards until it intersects the *LM*$_U$ line in point G. The goods market is not insulated from the money market disturbances.

7. Conclusion

A common currency has important benefits. In this chapter several of these benefits were identified. First, a common currency in Europe will eliminate transaction costs. This will not only produce direct, but also indirect benefits in that it will stimulate economic integration in Europe. Secondly, by reducing price uncertainty, a common currency will improve the allocative efficiency of the price mechanism. This will certainly improve welfare, although it is difficult to quantify this effect.

We have also concluded that we should not expect too much additional economic growth from a monetary union. The potential growth-boosting effects of a monetary union have been oversold. The theoretical reasons for a monetary union to stimulate economic growth are weak, and so is the empirical evidence. The benefits of a monetary union are to be found elsewhere than in its alleged growth-stimulating effects.

4

COSTS AND BENEFITS COMPARED

Introduction

In the previous chapters the costs and benefits of a monetary union were identified. In this chapter we conclude this discussion by comparing the benefits with the costs in a synthetic way. This will allow us to draw some general conclusions about the economic desirability of joining a monetary union by individual European countries.

1. Costs and benefits compared

It is useful to combine the figures (derived in the previous chapters) relating benefits and costs to the openness of a country. This is done in Fig. 4.1. The intersection point of the benefit and the cost lines determines the critical level of openness that makes it worthwhile for a country to join a monetary union with its trading partners. To the left of that point, the country is better off keeping its national currency. To the right it is better off when it relinquishes its national money and replaces it with the money of its trading partners.

Fig. 4.1 allows us to draw some qualitative conclusions concerning the importance of costs and benefits. The shape and the position of the cost schedule depends to a large extent on one's view about the effectiveness of the exchange rate instrument in correcting for the effects of different demand and cost developments between the countries involved.

At one extreme, there is a view, which will be called 'monetarist', claiming that exchange rate changes are ineffective as instruments to correct for these different developments between countries. And even if they are effective, the use of exchange rate policies typically make countries worse off. In this 'monetarist' view[1] the cost curve is very close to the origin. We represent this case in Fig. 4.2a. The critical point that makes it worthwhile to form a

---

[1] This is the view taken by the drafters of the influential EC Commission report (1990).

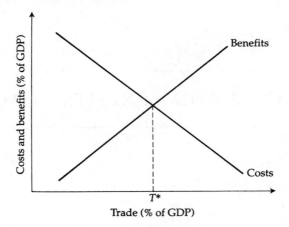

Fig. 4.1. Costs and benefits of a monetary union

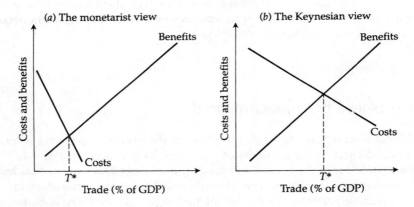

Fig. 4.2. Costs and benefits of a monetary union

union is close to the origin. Thus, in this view, many countries in the world would gain by relinquishing their national currencies, and by joining a monetary union.

At the other extreme, there is the 'Keynesian' view that the world is full of rigidities (wages and prices are rigid, labour is immobile), so that the exchange rate is a powerful instrument in eliminating disequilibria. This view is well represented by the original Mundell model discussed in Chapter 1. In this view, the cost curve is far away from the origin, so that relatively few countries should find it in their interest to join a monetary union. It also follows from this view that many large countries that now have one currency would be better off (economically) splitting the country into different monetary zones.

Table 4.1. Intra-Union exports and imports of EU countries (as per cent of GDP) in 1995

	Exports	Imports
Ireland	45.2	28.9
Belgium	43.9	39.9
Netherlands	32.2	24.9
Sweden	20.5	17.4
Finland	18.0	13.9
Portugal	17.4	22.8
Austria	15.5	19.7
Denmark	15.4	15.7
UK	12.1	12.5
Italy	11.9	11.2
Spain	11.8	13.3
Germany	11.8	9.9
France	11.6	11.3
Greece	5.3	14.5

Note: exports and imports include only goods, not services.
Source: EC Commission, *European Economy*, 62 (1996).

It is unmistakable that since the early 1980s the 'monetarist' view has gained adherents, and has changed the view many economists have about the desirability of a monetary union. Today the consensus seems to have evolved favouring monetary unification, certainly in Europe.

What does this analysis teach us about the issue of whether EU countries would benefit from a monetary union? In order to answer this question we first present some data on the importance of intra-EU trade for each EU country. The data are in Table 4.1, whose most striking feature is the large difference in openness of EU countries with the rest of the Union. This leads immediately to the conclusion that the cost-benefit calculus is likely to produce very different results for the different EU countries. For some countries with a large degree of openness relative to the other EU partners, the cost-benefit calculus is likely to tilt towards joining a European monetary union. This is most likely to be the case in the Benelux countries and Ireland.

It is surprising to find that Germany and France have the lowest degree of openness towards the rest of the EU (with the exception of Greece). Thus, if as is often said, France and Germany, together with the Benelux countries, form an optimum currency area, then other countries, including most Southern European countries, should also be added to this monetary area. (Note that this conclusion only considers one parameter, i.e. openness, in the cost–benefit analysis of a monetary union for the different countries. The other parameters, e.g. degree of flexibility or the degree of asymmetry in the shocks, may lead to a different conclusion. We will return to this issue.)

Note that some countries with a low trade share may nevertheless find it advantageous to join a monetary union. Our analysis of the credibility issues makes clear that high-inflation countries, like Italy, might decide that it is in their interest to join a monetary union despite the fact that the share of trade with the members of the union is relatively low. In terms of the analysis of Fig. 4.2, this implies that the Italian authorities do not consider the loss of the exchange rate instrument costly, so that the minimum trade share that makes the union advantageous is very low. Put differently, if one is sufficiently 'monetarist', one could argue that for countries with low degrees of openness, the benefits could still outweigh the costs, and joining a monetary union could also make sense for them from an economic point of view.

2. Monetary union, price and wage rigidities, and labour mobility

The cost-benefit calculus of a monetary union is also very much influenced by the degree of wage and price rigidities. As stressed in Chapter 1, countries in which the degree of wage and price rigidities is low experience lower costs when they move towards a monetary union. We show this in Fig. 4.3.

A decline in wage and price rigidities has the effect of shifting the cost-line in Fig. 4.3 downwards. As a result, the critical point at which it becomes advantageous for a country to relinquish its national currency is lowered. More countries become candidates for a monetary union.

In a similar way, an increase in the degree of mobility of labour shifts the cost curve to the left and makes a monetary union more attractive. In this sense it can be said that, if it increases labour mobility, the single market will make a monetary union more attractive for individual EU countries. It should be noted, however, that not all forms of integration have these effects. As

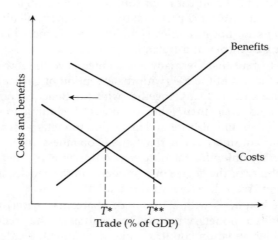

Fig. 4.3. Costs and benefits with decreasing rigidities

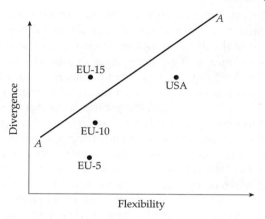

Fig. 4.4. Real divergence and labour market flexibility in monetary unions

stressed in Chapter 2, economic integration can also lead to more regional concentration of industrial activities. This feature of the integration process changes the cost-benefit calculus, in that it shifts the cost curve to the right and makes a monetary union less attractive.

3. Asymmetric shocks and labour market flexibility

Not only the degree of labour market flexibility (wage flexibility and labour mobility) matters for determining whether a monetary union will be attractive to countries. Also the size and the frequency of asymmetric shocks to which they are subjected matters. This means that countries that experience very different demand and supply shocks (because their industrial structures differ very much) will find it more costly to form a monetary union. In the framework of Fig. 4.3 this means that the cost-line shifts to the right.

We are now in a position to analyse the relation between labour market flexibility and asymmetric shocks in a monetary union. This is done graphically in the following way (see Fig. 4.4). On the vertical axis we set out the degree of 'real' divergence between regions (countries) which are candidates to form a monetary union. With real divergence is meant here the degree to which growth rates of output and employment tend to diverge as a result of asymmetric shocks.[2] On the horizontal axis we have the degree of flexibility

---

[2] The latter are those that occur independently from the monetary regime, and were described in Ch. 1. Asymmetric shocks that are the result of divergent national monetary policies are not included here. In a monetary union these would disappear.

of the labour markets in these regions (countries). The flexibility here relates to wage flexibility and interregional (international) mobility of labour.

The central insight of the theory of optimum currency areas is that countries or regions that experience a high divergence in output and employment growth need a lot of flexibility in their labour markets if they want to form a monetary union, and if they wish to avoid major adjustment problems. The larger is the degree of 'real divergence', the greater is the need for flexibility in the labour markets to make a smoothly functioning monetary union possible. This relationship between real divergence and flexibility is represented by the upward sloping line AA. Countries or regions located below the AA line can form a monetary union without 'excessive' adjustment costs. With excessive is meant here that the adjustment costs exceed the benefits of a monetary union. Countries above the AA line will experience a lot of adjustment costs if they form a monetary union. In other words, these countries have too low a degree of flexibility in their labour markets (given the level of real divergence). They do not form an optimum currency area. They are, therefore, well advised to maintain some degree of exchange rate flexibility. Of course, these countries are still free to form a monetary union. The theory, however, predicts that they will suffer economically from this decision.

Where should the European Union be located in Fig. 4.4? There is now a broad consensus among economists, who have tried to implement the theory empirically, that the *EU-15 is not an optimum currency area*. (See Eichengreen (1990), Neumann and von Hagen (1991), Bayoumi and Eichengreen (1993), De Grauwe and Heens (1993), De Grauwe and Vanhaverbeke (1993).)[3] Thus, according to these empirical studies, the EU as a whole (EU-15) is located above the AA line. As a result, from an economic point of view, a monetary union involving all EU member countries is a bad idea. The economic costs of a monetary union are likely to be larger than the benefits for a significant number of countries.

Whereas there is a strong consensus among economists that the EU-15 should not form a monetary union, there is an equally strong conviction that *there is a subset of EU countries which form an optimum currency area*. The minimum set of countries that could form a monetary union is generally believed to include Germany, the Benelux countries, and France (EU-5). This conclusion is buttressed by the same empirical studies as those quoted earlier.

Recent empirical analysis, however, has tended to enlarge the group of EU countries that would benefit from monetary union. The study of Artis and Zhang (1995) shows that the business cycles of EU countries (including Southern European countries) has become much more correlated since the early 1980s. Thus, there are now fewer asymmetric shocks among a relatively large group of EU countries than 15 years ago.[4] This seems to confirm what we said

---

[3] A dissenting view is presented in EC Commission (1990). See also Gros and Thygesen (1992).

[4] The authors credit the European Monetary System (EMS) for this. The EMS has forced countries to follow similar monetary policies. In so doing it has reduced the possibility of following independent monetary policies. The latter are a major source of asymmetric shocks.

earlier, i.e. that with economic integration the occurrence of asymmetric shocks tends to diminish.

Other recent studies have cast further doubts on the core–periphery view of monetary integration in the EU. Erkel-Rousse and Melitz (1995) and Canzoneri *et al.* (1996) find that in most EU countries monetary policies are powerless to affect real variables like output and employment. Thus, even if EU countries are confronted with asymmetric shocks, their national monetary policy instruments cannot be used to deal with them effectively. As a result, the loss of these instruments for most of the EU countries is not very costly.[5]

Finally, another series of empirical studies has found that a large part of the asymmetric shocks in the EU countries occur at the sectoral level and not so much at the national level. Put differently, a large part of the changes in output and employment in a country are the result of different developments as between sectors (e.g. due to demand shifts or to differential technological changes). These shocks cannot be dealt with by exchange rate changes. (See Bini-Smaghi and Vori (1993), Bayoumi and Prassad (1995), and Gros (1996).)

Thus, the most recent empirical studies lead to more optimism concerning the question of how many EU countries would profit from monetary union. This number could be significantly larger than the five or six core countries usually mentioned in this context. These recent empirical studies, however, do not seem to undermine our conclusion that the EU-15 as a whole does not constitute an optimum currency area. There is no consensus, however, about the size of the subset of countries that would profit from monetary union. In Fig. 4.4 we have placed the core group of countries (EU-5), about which there is a relatively wide consensus, below the *AA* line. (Note that within these countries the degree of labour market flexibility is not higher than among the member countries of the EU-15. The empirical evidence seems to indicate that the degree of real divergence is lower.) There are other subsets of countries, however (e.g. EU-10 = EU-5 + Austria, Ireland, Portugal, Spain, Italy), that according to the more recent studies could also form an optimum currency area. Because of the many difficulties in quantifying the costs of a monetary union, it will remain difficult, however, to obtain clear-cut results in this area.

In Fig. 4.4 we have also located the USA below the *AA* line. We are, of course, not really sure that the USA forms an optimum currency area. We are, however, much less uncertain about the relative position of the EU and the USA. Note that we have placed the USA at about the same vertical level as the EU-15, expressing the fact that the degree of real divergence between regions in the USA is not much different from the real divergence observed between countries in the EU (see Krugman (1993) on this). The major difference between the USA and the EU seems to be the degree of flexibility of labour

---

[5] For a recent study of Portugal confirming this, see Costa (1996).

markets. Many empirical studies have recently been done documenting this difference in the degree of flexibility of the labour markets in the USA and in Europe. For example, there is ample evidence that real wages in Europe respond less to unemployment than in the USA.[6] Similarly, there is ample evidence that labour mobility is much higher within the USA than it is between member countries of the EU.

It should be stressed here that the analysis underlying Fig. 4.4 is based on the traditional theory of optimum currency areas. It does not deal with some of the problems we have discussed in Chapter 2. For example, countries may experience asymmetric shocks because their monetary policies are independent. Some of them may find it difficult, for reasons of credibility, to follow low-inflation policies. The formation of a monetary union may reduce these problems. Thus, part of the asymmetric shocks one observes today in Europe may be the result of the absence of a monetary union. As a result, one cannot really be sure that the EU-15 would not gain from a monetary union. Nevertheless, with the present state of our knowledge, it is not unreasonable to maintain our conclusion that the EU-15 is not an optimum currency area.

The challenge for the EU-15 is to move to the other side of the *AA* curve, i.e. to make a monetary union less costly. How can this be achieved? There are essentially two strategies. One is to reduce the degree of real divergence, the other is to increase the degree of flexibility.

The difficulty of the first strategy is that the degree of real divergence is to a large extent dependent on factors over which policy-makers have little influence. For example, the degree of industrial specialization matters in determining how important asymmetric shocks are. There is very little policy-makers can do, however, to change regional specialization patterns.

There is one area, however, where policy-makers can do something to reduce the degree of real divergence. This is in the field of political unification. We argued earlier that an important source of asymmetric shocks is the continued existence of nation-states with their independent spending and taxing policies, and with their own peculiarities as far as economic institutions are concerned. In order to reduce asymmetric shocks, more economic policy co-ordination and institutional streamlining will be necessary. (A special problem arises here: How should labour unions be organized in a monetary union? We take up this issue in Box 5.)

The other strategy for moving the whole of the EU to the other side of the *AA* curve consists in increasing the degree of flexibility of labour markets (real wages and/or labour mobility). This strategy implies a reform of labour market institutions. Although such reforms are difficult to implement, they are necessary if one wants to have a monetary union involving the whole of the European Union.

---

[6] See Grubb *et al.* (1983) and Bruno and Sachs (1985).

4. A case-study

In this section we present a case-study that vividly demonstrates the difference in the adjustment process following shocks that affect regions (or countries) differently.

During the early 1980s the industrial world was hit by a severe recession. This world-wide downturn of economic activity hit regions of the world very differently. In general, regions with an older industrial structure suffered more severely. We take two examples, Michigan in the USA, and Belgium in Europe. Both regions are of a comparable size. In Figs 4.5 and 4.6 we present the evolution of the unemployment rate in these two regions and compare them with the total USA and European unemployment.

We observe from these figures that unemployment increased significantly more in Michigan and Belgium than in the USA and in Europe, respectively. It should also be noted that the fact that the USA is much more integrated economically than Europe did not prevent large differentials in unemployment from emerging. In fact the differential development in unemployment rates appears to be even more pronounced in the USA than in Europe. We discussed this phenomenon in Chaper 2, where we argued that integration also leads to regional concentration. In the case of the USA, integration has also led to a large regional concentration of the automobile sector in Michigan. This sector was severely hit by the recession of the early 1980s. This also explains the intensity of the unemployment problem in Michigan during that period.

How did these two regions adjust? The nature of the adjustment was very different in the two cases. In the case of Michigan a significant part of the

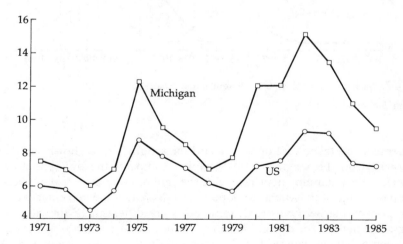

Fig. 4.5. Unemployment rate, Michigan and USA

Source: Eichengreen (1990).

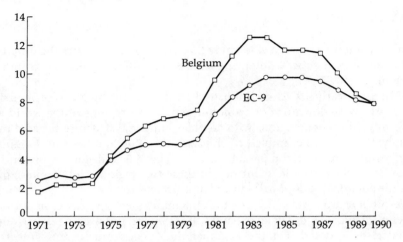

Fig. 4.6. Unemployment rate, Belgium and EC-9

Source: EC Commission (1990).

Fig. 4.7. Michigan unemployment differential and emigration

Source: Eichengreen (1990).

adjustment was taken care of by outward migration. This is shown in Fig. 4.7, borrowed from Eichengreen (1990), which compares the differential of the Michigan-US unemployment rate with the rate of emigration from Michigan (as a percentage of the Michigan population). Note that the percentages are not fully comparable because the denominators are different. In the case of the emigration figures, the denominator is total population, whereas in the case of the unemployment figures the denominator is the active population. As a result, a one per cent emigration rate involves more than twice the numbers implicit in a one per cent unemployment rate. It follows that emigration from

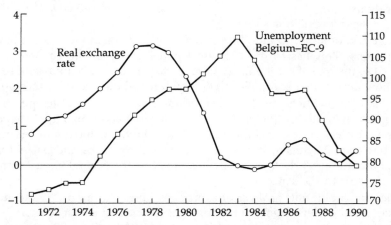

Fig. 4.8. Unemployment rate and real exchange rate, Belgium-EC

Source: EC Commission (1990).

Michigan was a sizeable fraction of the unemployed, and helped to reduce the unemployment rate of that state.

In the case of Belgium the adjustment process was mainly through the real exchange rate changes. In Fig. 4.8, we show the differential of the Belgian-EC unemployment rate together with the real exchange rate of the Belgian franc. We observe that Belgium adjusted to the unfavourable economic developments by a real depreciation of its currency of 20–25%. This helped to restore competitiveness, and started a process of gradual recovery leading to a significant narrowing of the unemployment differential between Belgium and the EC. At the end of the 1980s this differential had completely disappeared. About half of this real depreciation came about by nominal devaluations, the other half by lower cost and price developments in Belgium, compared to its main trading partners.

It should also be noted that the real depreciations of the early 1980s started to have effects on the unemployment rate in Belgium with some delay. This contrasts with the Michigan experience where the emigration reacted rather quickly to the worsening unemployment situation.

In the case of Michigan very little real depreciation took place. According to Eichengreen, regional changes in the real exchange rates in the USA were limited to a few percentage points during that period. (See Eichengreen (1990: 7). Note also that these real regional exchange rates in the USA can only change because of differential regional developments in prices.) In Belgium very little of the adjustment took the form of outward migration of Belgian unemployed workers. In 1984, for example, the number of Belgian workers who had emigrated to work elsewhere in the EC (i.e. the sum of the emigrations from all previous years) amounted to barely 40,000. This is 0.4% of the Belgian population.[7]

There is another important difference in the adjustment mechanism in these two regions. This has to do with fiscal policies. A separate chapter will be devoted to the role of fiscal policies in monetary unions. It is important to note here how differently fiscal policies work within and outside monetary unions.

In the case of Michigan the Federal budget tended to transfer purchasing power automatically to Michigan. This result came about mainly through the Federal transfers for the unemployed, and through the reduction in Federal tax revenues from Michigan. It has been estimated by Sachs and Sala-i-Martin (1989) that for every decline in state income of $1 the Federal budget transfers back 40 cents to the state.[8]

Belgium could not profit from such an intra-European redistribution. Instead, Belgium let its government budget deficit increase spectacularly and borrowed heavily in the foreign capital markets. This allowed it to soften the flow of the recession. It also implied that the *interregional* solidarity which was present in the USA was substituted for by an *intergenerational* solidarity within Belgium, where future generations of Belgians will have to service the government debt.

This case-study suggests that even a country like Belgium, for which the benefits of a monetary union probably outweigh the costs, would also find it costly to relinquish forever the exchange rate instrument in the face of large shocks such as the one that occurred in the early 1980s. If Belgium were to become a member of a monetary union, it would have to rely on alternative adjustment mechanisms. Since the mobility of labour between European countries is probably going to remain limited, these adjustment mechanisms are likely to be more painful than those that are initiated with exchange rate adjustments.

5. Costs and benefits in the long run

In the previous sections we discussed the costs and benefits of a monetary union. Our analysis was mostly static. It will be useful to add some dynamics to this analysis so as to obtain a better insight into the question of how these costs and benefits of monetary union may evolve over time.

In order to analyse this question, we will use some of the tools introduced in Chapter 2, where we discussed the relationship between the degree of economic integration and the occurrence of asymmetric shocks. This relationship

---

[7]   See Straubhaar (1988).

[8]   Von Hagen (1991) has argued that this number overestimates the automatic fiscal transfers in the USA.

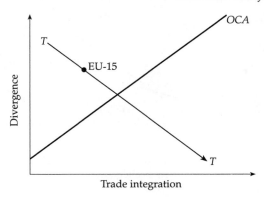

Fig. 4.9. The European Commission view of monetary integration

predicts whether the progress towards economic integration leads to economic convergence. We will now add the cost–benefit analysis.

In Fig. 4.9 we represent the model. On the vertical axis, we set out, as before, the degree of economic divergence between groups of countries. On the horizontal axis we have the degree of trade integration between the same groups of countries. The downward sloping line (*TT*) says that as trade integration increases the degree of economic divergence between the countries involved declines, i.e. countries become more alike and face less asymmetric shocks. (We called this the European Commission view in Chapter 2. We will return to the Krugman view later.) The upward sloping line (called *OCA*) represents the combinations of divergence and trade integration that make monetary union a break-even operation (costs = benefits). It is derived as follows. As trade integration increases, the net gains of a monetary union increase (see the previous sections). At the same time, when economic divergence increases, the costs of a monetary union increase. The two phenomena together allow us to derive the *OCA* schedule: when trade integration increases the net gains of the monetary union rise. These gains will be compensated by an increase in economic divergence. All points on the *OCA* line are then combinations of divergence and integration for which the monetary union has a zero net gain. Note that all the points to the right of the *OCA* line are points for which the benefits of monetary union exceed the costs. We call it the *OCA* zone.

In Fig. 4.9 we have put the EU-15 on the downward sloping *TT* line to the left of the *OCA* line, reflecting the conclusion that we arrived at in section 3, i.e. that today (in 1997) the EU-15 is probably not an optimum currency area. As trade integration within the EU proceeds, however, this point will move downwards along the *TT* line. This will inevitably bring us into the *OCA* zone, at least if we can assume that the dynamics of integration will continue to work within the EU. Thus, in this view, monetary unification will over

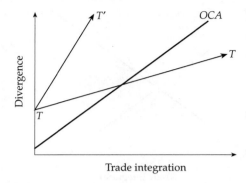

Fig. 4.10. The Krugman view of monetary integration

time be perceived to be beneficial for all countries in the European Union. In this sense monetary union is inevitable.

This is the optimistic view about the long-term prospects for monetary integration in Europe. There is, however, also a pessimistic view which one can derive from the Krugman analysis. It will be remembered that in the Krugman analysis, economic integration leads to more economic divergence between countries. This is represented in Fig. 4.10 by the upward sloping *TT* and *TT'* lines.

We now have to consider two possibilities for the long-term prospects of monetary union. One is represented by the *TT* line, the slope of which is flatter than the slope of the *OCA* line. In this case, although today the EU-15 may not be an optimum currency area, it will move into the *OCA* zone over time. In this case more integration leads to more specialization and thus more asymmetric shocks. However, the benefits of a monetary union also increase steeply with the degree of integration. As a result, despite the increase in asymmetric shocks, more integration will lead us into the *OCA* zone.

The second case is represented by the steep *TT'* line. Here integration brings us increasingly farther away from the *OCA* zone. This is so because the net gains of a monetary union do not increase fast enough with the degree of integration. As a result, the costs of divergence overwhelm all the other benefits a monetary union may have. In the long run the prospects for a monetary union of the EU-15 are poor.

From the discussion of the Krugman model we conclude that even if integration leads to more asymmetric shocks, this may still lead to increasing net gains of a monetary union for the EU-15. We cannot exclude, however, the possibility that the process of integration will make monetary union for the EU-15 more and more unattractive.

The reader should keep in mind what we said earlier about the relevance of the Krugman scenario. First, the theoretical presumption is not in favour

Box 5. Labour unions and monetary union

During our discussion of the costs and benefits of a monetary union we have stressed, on several occasions, the role of labour unions in determining the costs of a monetary union. Let us bring these insights together.

We established two rather opposite requirements for the optimal organization of labour unions in a monetary union.

First, we stressed that in the presence of *asymmetric* shocks, wages should be flexible, i.e. they should have different rates of change as between countries (and regions). An example we gave is a differential in productivity growth between countries. In that case nominal wage growth should be different and should reflect the differences in productivity growth between these countries (see Chapter 2). This implies that centralized wage bargaining would be harmful. By imposing the same nominal growth rates of wages it would lead to great losses of competitiveness of the country (region) with a low growth of labour productivity. This problem exists today within individual countries. A recent example is Germany where labour unions imposed centralized wage bargaining on the former East Germany. As a result, many East German firms failed to survive the shock of unification and employment was negatively affected. A similar problem exists in Italy where centralized wage bargaining has hurt the South of the country, where productivity is growing slower than in the North. The effect of this is that unemployment in the South of Italy is four to five times higher than in the North.

A second insight we obtained is that in the presence of the same, *symmetric* shock different wage bargaining systems may lead to a different wage–price spiral and therefore to divergent developments in competitiveness between countries (see Chapter 2). This seems to suggest that when symmetric shocks prevail the wage bargaining systems should be made more uniform across countries. Does this mean that a centralized wage bargaining system at the level of the union becomes desirable? Most probably not. The reasons are the following. First, we have noted that although in a unified Europe asymmetric shocks *between countries* may become less prevalent, specialization would still lead to large *regional* divergencies. These regional asymmetries may even increase in a more unified Europe. The characteristic feature of this specialization is that it would likely cross borders. In such an environment, a centralized wage bargaining system would be harmful for many of the European regions. Second, we have also stressed that technological changes tend to lead to uneven changes in output and employment between sectors. There is even a lot of empirical evidence indicating that many of the asymmetric shocks occur within the same sectors (see Davis *et al.* (1996)). Again, a centralized wage bargaining system would be very detrimental to output and employment in sectors that experience less favourable developments and for firms that lag behind other ones in the same sectors.

To conclude, the future organization of labour unions in a monetary union will have to respect the inevitable requirements of flexibility in a world where shocks occur mostly at the sectoral and micro-economic level.

of this scenario. Second, recent empirical evidence seems to sustain this theoretical presumption.

A last point about the long-run dynamics of monetary integration is the following. The decision by an individual country to join the monetary union is likely to speed up the integration process. In Chapter 2 we documented this by referring to empirical studies indicating how the existence of different national currencies helps to segment national markets. A decision by an individual country to join EMU, even if it does not qualify the *OCA* criteria, would have a self-fulfilling character, at least if Fig. 4.9 is the right view of the world. In this case the process of integration would be sped up by the very decision to join the monetary union, so that this new country grouping moves faster into the *OCA* zone.

6. Conclusion

The arguments developed in this chapter have led to the following conclusions. First, it is unlikely that the EU as a whole constitutes an optimal monetary union. Put differently, not all EU countries have the same interest in relinquishing their national currencies and in adhering to a European monetary union. The cost-benefit analysis of this chapter therefore also implies that a monetary unification in Europe will better suit the economic interests of the different individual countries if it can proceed with different speeds, i.e. if some countries who today feel that it is not in their national interest to do so, have the option to wait before joining the union.

Second, the number of countries that would benefit from monetary union is probably larger than what most economists thought just a few years ago. In addition, as the process of integration moves on, the number of countries that are likely to benefit from monetary union will increase. Thus, in the long run monetary union will be an attractive proposition for most, if not all, EU countries.

Third, even the countries that are net gainers from a monetary union take a risk by joining the union. The risk is that when large shocks occur (like the one that occurred in some countries in the early 1980s), they will find it more difficult to adjust, having relinquished their national currencies.

The discussion of this and the previous chapters has been based on an *economic* cost-benefit analysis. Countries may also decide to adopt a common currency for *political* reasons. A common currency may be the first step towards a political union that they wish to achieve. The economic cost-benefit analysis remains useful, however, because it gives an idea of the price some countries will have to pay to achieve these political objectives.

PART II
MONETARY
INTEGRATION

5

INCOMPLETE MONETARY UNIONS: THE EUROPEAN MONETARY SYSTEM

Introduction

In the previous chapters we discussed the costs and benefits of full monetary unions. In the real world, there exist many monetary arrangements between nations that are far removed from full monetary union, and yet also follow rules, and constrain the national monetary policies of the participants. The best-known 'incomplete' monetary union today undoubtedly is the European Monetary System (EMS), which was instituted in 1979 among a group of EC countries. Although in a legal sense it is still in existence, for all practical purposes it was abandoned in August 1993.

In this chapter, we analyse some of the operational problems of incomplete monetary unions, and in particular of the EMS (see Box 6 for a discussion of some basic institutional features of the EMS as it functioned during 1979–93).

Pegged exchange rate systems face some important problems that have led many economists to doubt the long-run sustainability of these systems. A first problem has to do with the credibility of the fixed exchange rates. A second one concerns the way the system-wide monetary policy is determined. The analysis of the problems will allow us to understand why the EMS experienced difficulties in 1992–3 which led to its disintegration.

1. The credibility problem

The credibility problem is closely linked to the analysis of the costs and benefits of relinquishing the exchange rate instrument. It arises for two different reasons. The first reason is that the use of the exchange rate can sometimes be the least-cost instrument to adjust the economy after some disturbance. The second reason follows from the Barro-Gordon analysis stressing that governments have different reputations, which may undermine the credibility of a fixed exchange rate. Let us analyse these two problems consecutively.

Box 6. The European Monetary System: some institutional features

The European Monetary System was instituted in 1979. It came as a reaction to the large exchange rate variablility of Community currencies during the 1970s, which was seen as endangering the integration process in Europe.[1]

The EMS consists of two features. The 'Exchange Rate Mechanism' (ERM) and the ECU. There is no doubt that the ERM has become the most important feature of the EMS.

Like the Bretton Woods system, the ERM can be called an 'adjustable peg' system. That is, countries participating in the ERM determine an official exchange rate (central rate) for all their currencies, and a band around these central rates within which the exchange rates can fluctuate freely. This band was set at +2.25% and −2.25% around the central rate for most member countries (Belgium, Denmark, France, Germany, Ireland, and the Netherlands). Italy was allowed to use a larger band of fluctuation (+6% and −6%) until 1990 when it decided to use the narrower band. The three newcomers to the system, Spain (1989), the UK (1990), and Portugal (1992), used the wider band of fluctuation. The UK dropped out of the system in September 1992. In August 1993 the band of fluctuation was raised to +15% and −15%.

When the limits of this band (the margins) are reached, the central banks of the currencies involved are committed to intervene so as to maintain the exchange rate within the band. (This intervention is called 'marginal' intervention, i.e. intervention at the margins of the band). The commitment to intervene at the margins, however, is not absolute. Countries can (after consultation with the other members of the system) decide to change the parity rates of their currency (a realignment).

These realignments were very frequent during the first half of the 1980s, when more than ten realignments took place. They have become much less frequent since the middle of the 1980s. During 1987–92 no realignment took place. In 1992–3 several realignments occurred. (In Section 8 we discuss post-1993 developments in the EMS.)

Countries can also undertake *intra-marginal* interventions, i.e. they can intervene in the market when the exchange rate has not yet reached one of its margins.

In order to undertake marginal interventions, the central banks need each other's currencies. For example, when the franc reaches its lower margin against the mark, the Banque de France will have to intervene by selling marks and buying francs. In order to sell marks, the Banque de France can have recourse to the 'Very Short-Term Financing Facility' (VSTFF). This is a system of credit lines between the central banks of the system. In our example, it allows the Banque de France to obtain an unlimited amount of marks which it can then sell in the foreign exchange market. In so doing, France becomes a debtor relative to Germany. It will have to repay the debt within 75 days after the end of the month in which the intervention took place.[2] Extensions of up to three months, however, are possible.

---

[1] For a fascinating account of the discussions that led to the establishment of the EMS, see Ludlow (1982). For a more detailed description of some institutional features of the system see van Ypersele (1985).

[2] Until 1987, when the Basle-Nyborg agreement was reached, this was 45 days.

Intra-marginal interventions can be undertaken without having recourse to the VSTFF. Until 1987 it was the rule that the members of the ERM would use dollars to do intra-marginal interventions. Thus, when France wanted to limit the downward slide of the franc within the band, it would buy francs and sell dollars. In 1987 the Basle-Nyborg agreement was signed among the members of the system allowing them to use the VSTFF in order to obtain a member country's currency for intra-marginal interventions.

The second feature of the EMS is the existence of the ECU. The ECU is defined as a basket of currencies of the countries that are members of the EMS. (This is a larger group of countries than the ERM members. It includes all the EU countries except Austria, Finland, and Sweden.)

The value of the ECU in terms of currency i (the ECU rate of currency i) is defined as follows:

$$ECU_i = \Sigma_j \, a_j \, S_{ji} \qquad\qquad (B6.1)$$

where a_j is the amount of currency j in the basket; S_{ji} is the price of currency j in units of currency i (the bilateral exchange rate).

In Table B6.1 we present an example of how the value of the ECU is computed in terms of French francs, using formula (B6.1). The first column shows the a_j's. Thus, we observe that one ECU contains 1.33 French francs, 3.43 Belgian francs, 0.198 Danish kroner, 0.624 German marks, etc. In the second column we have presented the exchange rates of EMS currencies in terms of the French franc. The third column is the product of the previous two columns. Each entry is the franc-equivalent of the amount of currency j in the basket. By summing this column one obtains the price of the ECU in units of francs.

Table B6.1. The valuation of the ECU in units of French francs

Currency	a_j (1)	S_{ji} (2)	$a_j S_{ji}$ (3)
French franc	1.33	1	1.33
Belgian franc	3.43	0.164	0.563
Danish krone	0.198	0.882	0.175
German mark	0.624	3.378	2.108
Irish pound	0.0085	8.330	0.071
Italian lira	151.8	0.003	0.514
Dutch guilder	0.2198	3.012	0.662
Pound sterling	0.0878	8.284	0.727
Greek drachma	1.44	0.022	0.031
Spanish peseta	6.885	0.040	0.277
Portuguese escudo	1.393	0.034	0.047
		$\sum_j a_j {}^* S_{ji} =$	6.924

Source: computed from *Financieel Economische Tijd*, 22 October 1996.

In order to obtain the other ECU rates one can repeat the previous exercise after substituting in column (2) the exchange rates of the EMS currencies in units of the relevant currency (say currency k). Alternatively, one can use the triangular arbitrage condition, i.e.

$$ECU_k = ECU_i.S_{ik} \tag{B6.2}$$

where ECU_k = the price of the ECU in terms of currency k. Both procedures should yield the same result.

It should be noted that the ECU markets have evolved in such a manner that the daily ECU exchange rates are now determined by demand and supply. The ECU rates computed by equation (B6.1) are therefore theoretical values. Arbitrage should ensure that the market rates of the ECU cannot deviate too much from its theoretical rates. During the last few years, however, relatively large differences between the market and the theoretical rates of the ECU have emerged. This suggests that the scope for arbitrage has been reduced.

The ECU has a number of peculiar features. We analyse two of these here. First, currencies that tend to depreciate in the exchange market against the other currencies in the basket will see their share in the basket decline. This can be shown as follows. In Table B6.1 we observe that the share of the French franc in the ECU can easily be defined as the amount of francs in the basket (1.33) divided by the French franc value of the ECU (the sum of the last column). More generally the share of currency i in the ECU is

$$b_i = a_i/ECU_i \tag{B6.3}$$

It can now easily be seen that, as currency i depreciates against all the currencies of the basket, it will also depreciate against the ECU. In other words, the price of the ECU in units of currency i (ECU_i) increases. Since a_i is constant, b_i in equation (B6.3) must decline.

This feature of the ECU by which depreciating currencies experience a decline of their weight in the basket (and appreciating currencies experience an increase of their weight) has led to problems. If the a_i's (the *amount* of currency i in the basket) are left unchanged, the strong EMS currencies will continuously increase in importance in the valuation of the ECU. For political reasons it was felt that this was unacceptable. As a result it was decided that every five years the a_i's are changed so as to maintain shares that are relatively stable in the long run. These changes typically implied that the amounts of the weak currencies were increased and those of the strong currencies were lowered. Many experts are convinced that this was a very unattractive feature of the ECU, since it introduced uncertainty about the future value of the ECU. In the Treaty of Maastricht (1991) it was agreed to solve this problem by freezing the a_i's.

A second feature of the valuation of the ECU is that when a currency depreciates (appreciates) against the other currencies in the basket, the depreciation (appreciation) against the ECU will typically be lower. This can again be seen from Table B6.1. Suppose that the French franc depreciates by 10% against all the other EMS currencies. In column (2) we now have to multiply all numbers by 1.1, except the French entry because the franc cannot depreciate against itself. Thus in the third column we still have the same entry for France. This means that the price of the ECU in terms of francs (the sum of the last column) will increase by less than 10%. The larger the share of the franc the lower its depreciation against the ECU.

What is the role of the ECU in the EMS?

The founding fathers of the EMS had high hopes about the role of the ECU in the EMS. Until today, however, the ECU has only played a limited role in the EMS. In fact, it is no exaggeration to say that the exchange rate mechanism of the EMS functioned without the ECU.

The ECU is used as a unit of account in the EMS in the following way. Central rates are officially defined in terms of ECU rates. These central ECU rates are then translated into bilateral central rates using the triangular arbitrage condition, i.e.

$$S^*_{ij} = ECU^*_j/ECU^*_i \tag{B6.4}$$

where S^*_{ij} is the central rate of currency i in units of currency j; ECU^*_i and ECU^*_j are the central ECU rates of currencies i and j respectively.

This procedure then yields the so-called parity grid. It is clear that only the bilateral central rates have operational significance, because the interventions in the foreign exchange markets are carried out in national currencies and not in ECU.

The founding fathers of the EMS also wanted the ECU to be used as an instrument which would bring more symmetry into the system of interventions in the exchange markets. Therefore an indicator of divergence was defined based on the ECU. When a currency's ECU rate would diverge by more than 75% of its permitted band of fluctuation, the country in question was supposed to undertake measures to correct the divergence. This indicator of divergence was intended to work symmetrically, i.e. it would indicate which currency was on average becoming weaker or stronger against the other currencies.

This indicator of divergence never worked very well, and was quickly abandoned. The EMS did not become the symmetric system it was supposed to be. Instead relatively strong asymmetries emerged. The reasons why this happened are analysed in the main text.

1.1 Adjustment problems lower the credibility of a fixed exchange rate

Let us use the framework of Chapter 1 to analyse the adjustment problem. We repeat the figure of Chapter 1 here (see Fig. 5.1). Suppose that the country (called France) has committed itself to keeping its exchange rate fixed. There now occurs a wage explosion in France (say like the one that happened after the 'events' of May 1968). We represent this shock by an upward movement of the supply curve of France. The effect of this shock is to lead to a current account deficit and a reduction of French output and employment. As stressed earlier in Chapter 1, if country A's wages and prices are flexible, there is nothing to worry about: French wages will decline so that the supply curve shifts downwards again. Similarly, if French workers are willing to emigrate, the unemployment problem disappears. If, however, wages are rigid, and/or if workers are immobile, France faces a dilemma situation: it can reduce the

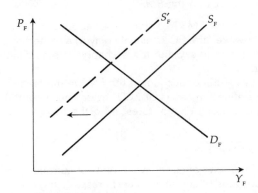

Fig. 5.1. The dilemma problem of France

current account deficit by following deflationary monetary and fiscal policies, at the cost of a still lower output and employment level; or it can increase output by expanding monetary and fiscal policies at the cost of a higher current account deficit.

With its commitment to fixed exchange rates (and given wage rigidity) there is no way France can resolve this dilemma. Note that this dilemma arises here because France pursues two objectives, while it has only one instrument, i.e. general demand policies. It has, however, one other instrument which it has chosen not to use, i.e. the exchange rate instrument. It is clear that the incentive to use this instrument in order to resolve the dilemma will be great. Speculators also will realize that the government has this incentive. As a result, they will believe it likely that a devaluation will occur in the future.

Thus, when two or more countries decide to peg their exchange rates, economic agents realize that there may be future situations in which a change in the exchange rate would minimize adjustment costs. They therefore also know that when these future situations occur, the authorities will have an incentive to renege on their commitment to keep the exchange rate fixed. As a result, such a fixed exchange rate commitment will have a low credibility, and will often be subjected to speculative crises. This credibility problem can be solved only if the authorities convince speculators that their only objective is the maintenance of the fixity of the exchange rate, whatever the output and employment costs.

1.2. Differences in reputation lead to low credibility of a fixed exchange rate

The Barro-Gordon model, which we discussed in Chapter 2, allows us to point to a second reason why a pegged exchange rate system suffers from a credibility problem. This analysis led to the conclusion that the high-inflation country (Italy) has a lot to gain from pegging its currency to the currency of a low-inflation country. However, Italy will find it difficult to fix its exchange rate credibly. Once the exchange rate is fixed, it has an incentive to follow inflationary policies and to devalue by surprise, so as to obtain a more favourable inflation-unemployment outcome. This, however, will be anticipated by economic agents, who will adjust their expectation of inflation. Italian inflation will be permanently higher than in Germany. This will force the authorities to devalue the currency regularly. The fixity of the exchange rate has no more credibility than announcing a low inflation rate if the Italian authorities are perceived as 'wet' and short-sighted. In addition, the regular devaluations will be anticipated, leading to large-scale speculative crises prior to the anticipated dates of realignments. The fixed exchange rate commitment will collapse. In Section 5 we return to these issues when we analyse the reasons for the collapse of the EMS in 1993.

2. The liquidity problem in pegged exchange rate systems

Every system of fixed exchange rates faces the problem of how to set the system-wide level of the money stock and interest rate. This issue essentially arises from the so-called $n - 1$ problem. In a system of n countries, there are only $n - 1$ exchange rates. Therefore $n - 1$ monetary authorities will be forced to adjust their monetary policy instrument so as to maintain a fixed exchange rate. There will be one monetary authority which is free to set its monetary policy independently. Thus, the system has one degree of freedom. This leads to the problem of how this degree of freedom will be used. Who will be the central bank that uses this degree of freedom? Are there alternatives to the asymmetric solution where one central bank does what it likes, and others follow?

These are the questions analysed in this section. We do this by using a very simple two-country model of the money markets. Let us represent the money market of country A by the following equations:

money demand: $M_A^D = P_A L_A(Y_A, r_A)$ (5.1)

money supply: $M_A^s = R_A + D_A$ (5.2)

The money demand equation is specified in the traditional way. We assume that an increase in the price level of country A (P_A) increases the demand for money. Similarly an increase in output, Y_A, raises the demand for money. A reduction in the interest rate r_A, increases the demand for money.

The money supply consists of two components, the international reserve component, R_A, and the domestic component of the money supply, D_A (the latter consists of bank credit to the domestic private and government sectors).

For country B we postulate similar equations, i.e.

money demand: $M_B^D = P_B L_B(Y_B, r_B)$ (5.3)

money supply: $M_B^S = R_B + D_B$ (5.4)

We will assume that there is perfect mobility of capital between these two countries. This allows us to use the interest-parity condition, which we specify as follows:

$$r_A = r_B + \mu$$ (5.5)

where μ is the expected rate of depreciation of the currency of country A.[3]

This relationship is also called the 'open interest parity'. It says that if economic agents expect a depreciation of currency A, the interest rate of country A will have to exceed the interest rate of country B in order to compensate holders of assets of country A for the expected loss.

Now suppose countries A and B decide to fix their exchange rate. Let us also assume that economic agents do not expect that the exchange rate will be adjusted in the future. This means that $\mu = 0$. The interest rates in the two countries will be identical.

We can now represent the equilibrium of this system graphically as follows (Fig. 5.2). The downward-sloping curve is the money demand curve. The money supply is represented by the vertical lines M_A^1 and M_B^1. Money market equilibrium in both countries is obtained where demand and supply intersect (points E and F). In addition, given the interest-parity condition, the interest rates must be equal.

It is clear from Fig. 5.2 that there are many combinations of such points that bring about equilibrium in this system. Consider, for example, the points G and H. At these two points demand and supply of money in the two countries are equal, and the interest rates are also equalized. It is easy to see that there are infinitely many combinations that will satisfy these equilibrium conditions. Each of these combinations will produce one level of the interest rate and one of the money stock. One can say that the fixed exchange rate arrangement is compatible with any possible *level* of the interest rates and of the money stocks. There is a fundamental indeterminacy in this system. This follows from the $n - 1$ problem, which, as we have seen, produces one degree of freedom in the system.

How can this indeterminacy be solved? We discuss two possible solutions: one is asymmetric (hegemonic), the other is symmetric (co-operative).

---

[3] In Box 7 more explanation is given about how this expression is derived.

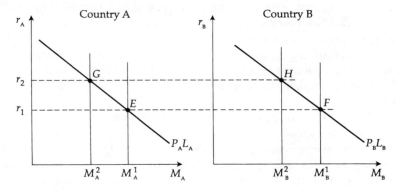

Fig. 5.2. The *n* − 1 problem in a two-country monetary model

2.1 The asymmetric (hegemonic) solution

The first solution to the problem consists in allowing one country to take a leadership role. Suppose, for example, that country A is the leader and that it fixes its money stock independently, say at the level M_A^1 (see Fig. 5.2). This then fixes the interest rate in country A at the level r_1. Country B now has no choice any more. Its interest rate will have to be the same as in country A. Given the money demand in country B, this then uniquely determines the money supply in country B (M_B^1) that will be needed to have equilibrium. Country B has to accept this money supply. It cannot follow an independent monetary policy.

In this asymmetric arrangement, country A takes on the role of anchoring the money stock in the system. The degree of freedom in the system is used by country A to set its monetary policy independently.

2.2 The symmetric (co-operative) solution

A second possibility is for the two countries to decide jointly about the level of their money stocks and interest rates. Thus, this solution requires co-operation. For example, in Fig. 5.2, the two countries could decide jointly that the money stocks in the two countries will be equal to M_A^1 and M_B^1.

The original blueprint of the EMS was aimed at promoting this co-operative solution. In particular, the use of the ECU as an indicator of divergence was seen as an instrument that would promote symmetry in the system. As explained in Box 6, when the market exchange rates of a currency deviated too much from its central rates, this would show in the indicator of divergence. The country involved would then be singled out to take necessary action. This implied that a country with a strong (weak) currency would be required to expand (contract) its monetary policies.

A second way the symmetric solution was intended to work was through the system of interventions in the foreign exchange markets. The rule was that when two currencies hit their upper limits, the intervention would be in each other's currency, so that the monetary effect in the two countries would be symmetrical.

We illustrate this symmetric intervention system in Fig. 5.3. We assume that doubts have arisen about the fixity of the exchange rate of currency A against currency B, and that economic agents expect a *future* devaluation of currency B. According to the interest-parity condition, this requires an increase in the interest rate of country B relative to the interest rate of country A. In the foreign exchange market the following will happen. The expectation of a future devaluation of currency B leads speculators to sell currency B against currency A. In order to prevent the market rate of currency B from dropping below its limit against A, the central bank of country B must buy its own currency and sell currency A. (The latter it will typically obtain through the system of short-term financing, which forces country A to provide the necessary amounts of its currency to country B.) The result of this intervention on the money stocks is symmetric: country B's money stock declines, and country A's money stock increases. The latter arises from the fact that the sale of currency A by country B increases the amount of currency A in circulation. We represent this symmetric effect in Fig. 5.3 by a leftward shift of the money stock line in country B (from M_B^1 to M_B^2) and a rightward shift of the money stock line in country A (from M_A^1 to M_A^2). As a result, the interest rate in country B increases to r'_2, whereas it declines in country A to r_2. Thus, the speculative disturbance is taken care of by a symmetric adjustment, in which country A allows its money stock to increase and its interest rate to decline, and country B allows the opposite to occur.

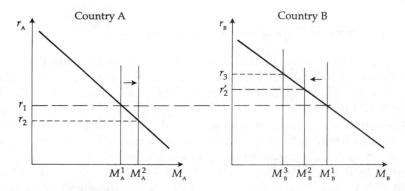

Fig. 5.3. Symmetric and asymmetric adjustment to a speculative movement

As a matter of fact, the symmetric solution, as we just described it, has generally not worked well. In particular, when a speculative crisis arose, requiring intervention in the foreign exchange market, the strong-currency country (Germany) has generally been unwilling to allow its money stock to increase and its interest rate to decline. This Germany has achieved by using *sterilization policies*, i.e. by offsetting the expansionary effects of the interventions in the foreign exchange market by reverse operations of the Bundesbank. Thus, when the central bank of the weak-currency country sold marks (against its own currency), the Bundesbank has usually bought these marks back through open market operations. The effect of these sterilization policies then typically was that the German money stock was not (or only slightly) affected by the foreign exchange market operations of the weak-currency countries.

The implication of this asymmetry is that the weak currency is forced to do all the monetary adjustment. In Fig. 5.3 we see that country B is now forced to reduce its money stock to the level given by M_B^3, and to allow the interest rate to increase to r_3. The money stock of country A remains unchanged at M_A^1 in this asymmetric adjustment system.

Despite the original intentions of the founding fathers of the EMS, the system has evolved into an asymmetric one.[4] Such asymmetric arrangements are very common in fixed exchange rate systems. In the Bretton Woods system, the USA took on this role. In the EMS, Germany has taken the anchoring role.

Several issues arise here. First, the question is why a particular country is selected as a leader, and not another one. Why did Germany become the leader in the EMS? This question will be analysed in Section 2.4 below. Secondly, what are the characteristics of these asymmetric (sometimes also called hegemonic) arrangements as compared to symmetric systems, where countries decide jointly about the system-wide money stock and interest rate, or where they allow rules to lead to symmetric adjustments?

2.3 Symmetric and asymmetric systems compared

An asymmetric system has a number of important advantages, together with disadvantages. Let us analyse the advantages first. One important advantage of this system is that it imposes a lot of discipline on the peripheral country. Suppose the latter should decide to increase its money stock. This would immediately lead to a drain on the international reserves of the central bank of the peripheral country, because residents would seek a riskless higher rate of

---

[4] There has been a lively academic discussion on the question of how important the dominant position of Germany has been during the EMS period. Econometric analysis suggests that, although the influence of German monetary policies on monetary conditions in the other EMS countries has been pervasive, this influence is not one way. Other countries' monetary policies have occasionally also influenced German monetary conditions. For evidence, see Fratianni and von Hagen (1990) and De Grauwe (1991).

return in the centre country. The peripheral country would be forced almost instantaneously to lower its money stock again.

It should be noted here that in the asymmetric arrangement the flow of liquidity from the periphery to the centre country does not affect the money stock in the latter country. In this asymmetric system, the centre country automatically sterilizes the liquidity inflow by reverse open market operations. If it did not do so, the monetary expansion engineered by the peripheral country would lead to an increase of the money stock in the centre country. That country would then also lose its function of an anchor for the whole system.

There are other features of the asymmetric system, however, that make it less attractive. These have to do with the way unsynchronized business cycles affect the money markets. In order to show this, we suppose that the peripheral country experiences a recession. We represent this effect in the money markets of the two countries in Fig. 5.4. The recession in the peripheral country shifts the demand for money downwards as shown in Fig. 5.4. This has the effect of reducing the interest rate in the peripheral country. However, interest parity (and assuming that no future devaluations are expected) makes it impossible for the interest rate to decline below the centre country's interest rate. Since in the centre country no change in the money demand occurs, and its authorities continue to fix the domestic money supply, the interest rate in the centre country cannot change. The result is that the money supply in the peripheral country must automatically decline. This comes about as follows. The downward pressure on the peripheral country's interest rate leads to an outflow of capital to the centre country. This reduces the money stock in the peripheral country, and would increase the money stock in the centre country. However, as the authorities of that country are committed to fix their supply of money, they automatically sterilize the reserve inflow by open market sales of securities.

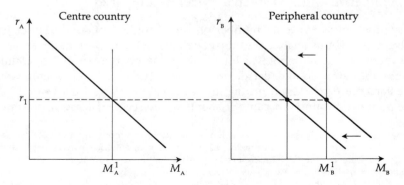

Fig. 5.4. The effects of a recession at the periphery

The less attractive feature of this adjustment mechanism is that a recession which originates in the peripheral country is made worse by a contraction of the country's money stock. The opposite occurs when a boom originates in the peripheral country. In that case the boom is automatically accommodated by an increase in the money stock of the peripheral country, because the upward pressure on its interest rate leads to a capital inflow. (The reader can work out this case for herself/himself using the model of Fig. 5.4.) The result is that the business cycles in the peripheral country are likely to be made more intense by the procyclical movements of the money stock of the periphery.

It is clear that this asymmetric system of monetary control is not very efficient at dealing with asymmetric shocks such as unsynchronized business cycles. The problem is that while the centre country tries to stabilize its money stock without regard to what happens in the rest of the system, it helps to make the money stock volatile in the peripheral country. In this system, there is no one responsible for the money stock of the whole system.

The previous analysis also makes clear that the asymmetric system in which the centre country rigidly fixes its money supply does not guarantee that the money stock will be stable in the system as a whole when asymmetric shocks occur. The symmetric system of monetary control would be more successful in stabilizing the system's money stock when unsynchronized business cycles occur. In order to show this, we take the case of a recession in the peripheral country again. We now assume that the central banks of the centre and the periphery co-operate to stabilize the whole system's money stock. They can achieve this as follows. The peripheral country reduces its money stock, and the centre country increases its money stock. We show this case in Fig. 5.5.

The effect of this co-ordinated approach is that the total money supply is kept unchanged. Note that this result will come about automatically if the centre country does not sterilize the inflows of reserves that are triggered by the decline in the interest rate of the peripheral country.

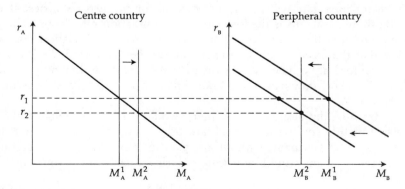

Fig. 5.5. A recession at the periphery in a symmetric system

The problems of monetary control that arise in an asymmetric system when asymmetric shocks occur are likely to lead to conflicts about the kind of monetary policy to be followed for the whole system. For example, during a recession originating in the peripheral countries, these will find that the monetary policies they have to endure are inappropriate. Pressures on the centre country will certainly be exerted. These conflicts of interests will have to be dealt with in one way or another.

The preceding discussion also suggests that an asymmetric system may not survive in the long run. Too much conflict will exist about the appropriate monetary policies for the system as a whole. Peripheral countries, especially if they are similar in size to the centre country (as is the case in the present EMS), may not be willing to subject their national interests to the survival of the system. In the end, more explicit co-operative arrangements may be necessary.

2.4. The choice of a leader in the EMS

In the previous sections it was argued that asymmetric arrangements are often used to 'anchor' the money stock in a fixed exchange rate system. In this section we analyse the question of why fixed exchange rate systems tend to end up with one leader and many followers. In addition, we determine the factors that decide who becomes the leader. In order to do so, we return to the Barro-Gordon model of two countries. Let us call these two countries Germany and Italy again.

We take as a starting point Fig. 2.14 (Chapter 2). Suppose, as before, that Germany and Italy have decided to fix their exchange rate. In principle, this arrangement can work with any level of the common inflation rate. This could be the German inflation rate. It could also be the Italian inflation rate, or any other inflation rate, as long as it is the same one for both countries. This requirement follows from the purchasing power parity condition which we have imposed.

Which inflation rate will be most beneficial for the two countries? It can easily be shown that this is the lower (German) inflation rate. In order to see this, suppose first that Italy decides to accept the German inflation rate, and suppose that the fixing of the exchange rate can be made credible. Then the welfare gain for Italy will be given by the movement of the inflation rate from E to F (see Fig. 5.6). Italy has now a lower inflation rate at no cost in terms of unemployment.[5]

Consider now the alternative. Germany would accept the Italian inflation rate. The inflation equilibrium would now be given by point G. This is a clear loss of welfare in Germany, without any gain in Italy. There is no reason to assume that Germany would be willing to strike a deal which would make it

---

[5] We abstract from the possible short-run losses when Italy disinflates and unemployment increases temporarily.

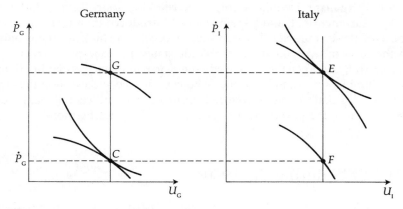

Fig. 5.6. The selection of a leader in the EMS

worse off, and which would not make Italy any better off. In fact, it can also be seen that Germany has no incentive at all to accept any other inflation rate than the one it had prior to the establishment of the fixed exchange rate. Conversely, Italy has an incentive to aim for the lower German inflation rate. Under those conditions, it is quite easy to understand that Italy will be glad to accept the leadership of Germany in determining the system-wide inflation rate. This is an arrangement that maximizes welfare in Italy without reducing welfare in Germany.

The previous analysis stresses that the leadership position of a country depends on its reputation in maintaining a low inflation equilibrium. Reputations can change, however. For example, political and institutional changes may affect the preferences of the centre country's authorities, leading to a higher inflation equilibrium. Similarly, an increase in the natural unemployment rate may bring about a higher inflation equilibrium. These changes are likely to jeopardize the leadership position of the centre country, as other countries may decide to stop pegging their exchange rates to that country, and to look for another low-inflation country to take up the role of the leader.

Such changes in leadership positions are likely to lead to great disturbances. They can also endanger the stability of the system, and may lead to its downfall. This is what happened in the Bretton Woods system when the USA stopped being the low-inflation country of the system. Other countries lost their incentives to peg their currencies to the dollar.

Something similar has happened recently in the newly independent states of the former USSR. In the early 1990s Russia increasingly moved into a situation of monetary instability and hyperinflation. The newly independent states had the choice either to stay in the rouble area, and thus to import hyperinflation, or to drop out of the rouble zone and to introduce their own currency which would float against the rouble. Many states decided to do the latter. At the end of 1993 Azerbaijan, Estonia, Georgia, Kyrgyz Republic, Latvia, Lithuania, Moldova, Turkmenistan, and Ukraine had all introduced their own currencies. In some of these countries this step was successful in shielding the economy from the Russian hyperinflation (the Baltic states, for example). Others were not successful, and in fact experienced even worse monetary instability than in Russia (Ukraine, for example). Those who experienced success were the countries where legislation was introduced ensuring that the central bank could refuse to finance the government budget deficits by printing money.

3. The EMS during the 1980s

The surprising fact about the EMS is that it lasted for more than a decade before it succumbed to the problems we identified in the previous sections. Thus, before we study the reasons why the system disintegrated in 1992–3, it will be helpful first to understand why it took thirteen years before it finally happened. What were the factors that helped the system temporarily to overcome the problems discussed in the previous sections? We identify two features of the early EMS, the bands of fluctuation and the existence of capital controls.

3.1. Bands of fluctuation

First, there is the existence of *bands of fluctuation*. Contrary to the Bretton Woods system, where they were relatively narrow (1% above and below the parities), in the EMS the bands were fixed at 2.25% above and below the official parities (central rates). This allowed exchange rates to move by (at most) 4.5%. For some countries (Italy until July 1990, Portugal, Spain, and the UK) the bands of fluctuation were even set at 6% below and above central rates, allowing a maximal range of fluctuation of 12%. These relatively large bands of fluctuation were quite important in stabilizing the system in the following way.

The existence of relatively wide bands made it possible for the authorities of high-inflation countries to change the exchange rate regularly by small amounts without having to face large speculative crises prior to each expected realignment. A country with high inflation (e.g. Italy which faced a band of 12%) could devalue its currency by, say, 8% per year to offset the larger inflation rate. Given the fact that the bands of fluctuations were 12%, after each realignment the new central rate would always fall between the previously prevailing upper and lower limits. As a result, after the realignment the market exchange rate most often would not change, and could even decline. We show this feature in Fig. 5.7. We suppose that prior to the realignment date, at t_1, the exchange rate (the price of the German mark in units of lira) hits the upper limit. At t_1 the lira is devalued against the mark (the official price of the mark

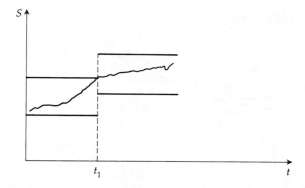

Fig. 5.7. Realignments and bands of fluctuation in the EMS

increases) by 8% to reflect a higher inflation of 8% in Italy as compared to Germany. One day later the previous market rate will fall between the new limits. The change that occurs on that day will usually be small. Quite often the market exchange rate will fall the day after the realignment, as speculators take their profits.

This feature of the working of the EMS is quite different from the Bretton Woods system, where the total band of fluctuation was only 2%. We show the contrast in Fig. 5.8. Suppose again that Italy experiences a rate of inflation that is 8% higher than in the USA. At time t_1 Italy decides to devalue by 8%. The day after, the new market rate will necessarily have to jump up by at least 6%. This feature creates huge profit opportunities. Speculators expecting the devaluation to occur on t_1 will be willing to bet very large amounts of money (i.e. to buy large amounts of dollars against lira) to profit from the expected gain. These speculations have a typically asymmetric feature: if the devaluation occurs on t_1 speculators make a large gain; if the devaluation does not happen they make only a small loss. The loss then arises from the fact that the interest rate on dollar assets (which have been bought) is usually smaller than the interest rate on lira assets (which have been sold). As these speculative movements are short term, this loss is very small. This one-way bet feature of speculation was very prevalent during the Bretton Woods system. It was most often absent in the EMS during the 1980s, mainly because monetary authorities were able to keep the size of the realignments small compared to the size of the bands of fluctuation. We show evidence for this in Fig. 5.9, which presents the realignments of the lira and the French franc together with the bands of fluctuation. It can be seen that during the 1980s all the realignments of the lira were smaller than the size of the band, making it possible that after each realignment the market rate of the lira moved smoothly, without jumps. This was not always the case with the French franc, however. We observe that during the period 1982–3, the three realignments of the French

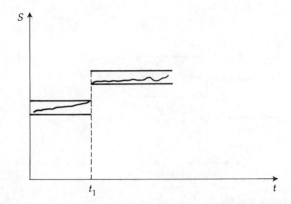

Fig. 5.8. Realignments and bands of fluctuation in the Bretton Woods system

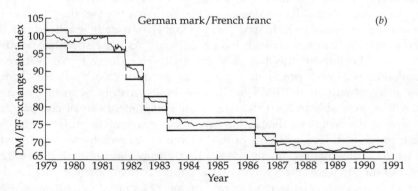

Fig. 5.9. Bilateral EMS exchange rates and bands of fluctuation

Source: EC Commission.

franc were significantly larger than the width of the band, creating large speculative gains. This period was also very turbulent. At some point, there was serious talk that the EMS would not survive the frequent speculative crises. After that period, however, the French authorities succeeded not only in reducing the frequency of the realignments but also in limiting the size of the realignments (relative to the size of the band). This certainly contributed to reducing the scope for speculative gains.

We conclude from the preceding discussion that the relatively large bands of fluctuations in the EMS together with relatively small and frequent realignments, while not preventing some countries from having larger rates of inflation than others, helped to reduce the size of speculative capital movements and stabilized the system during most of the 1980s. This feature of the early EMS changed drastically after 1987. From that date on, the system evolved into a much more rigid exchange rate arrangement. The EMS countries made it clear that their ambition was to keep the exchange rates fixed. Thus, the previously relatively flexible exchange rate arrangement evolved into a truly fixed exchange rate system. In this new environment, the problems identified in the previous section were to start playing an increasing role.

3.2. Capital controls

The early EMS was also characterized by the existence of capital controls. In particular, France and Italy maintained such controls during most of the 1980s.[6] These controls tended to reduce the size of funds that could be mobilized for attacking a currency. In so doing, they gave the authorities some time to organize an orderly realignment of the exchange rates. Thus, the function of capital controls was not to maintain unrealistic exchange rates. Rather it was a mechanism that, in conjunction with the willingness frequently to realign the exchange rates, allowed this process to occur with a minimum of turbulence.

Things changed at the end of the 1980s when France, and later Italy, gradually eliminated the controls on capital movements. This change was brought about mainly as a result of the decisions to move towards a fully integrated internal market in the European Community. The free movement of capital was seen as an essential part of complete market integration. Together with the movement towards much greater fixity of the exchange rates, this move changed the nature of the EMS.

4. The disintegration of the EMS in 1992–3

At the start of the 1990s the EMS had evolved into a truly fixed exchange rate system with (almost) perfect capital mobility. In this new monetary environ-

---

[6] Belgium had a system of dual exchange markets, separating the current and capital transactions.

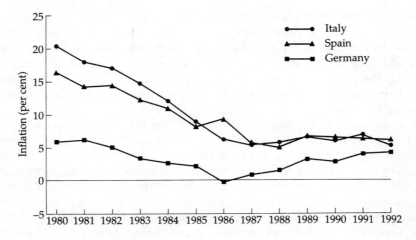

Fig. 5.10. Inflation rates of Italy, Spain, and Germany

Source: EC Commission (1993).

ment, the credibility and the liquidity problems identified in sections 1 and 2 started to have their full destabilizing effects.

4.1. The lira and the peseta crises

A first problem arose mainly in connection with the lira and the peseta. Despite vigorous and in a sense successful efforts at reducing inflation in Italy and Spain, these two countries did not manage to close the inflation gap with Germany. This certainly had something to do with differences in the reputation of the monetary authorities in Italy and Spain versus Germany which made it difficult to reduce inflationary expectations in these southern countries to the German level. As a result, actual inflation rates in these countries failed to move to the German level. We show the evidence for Italy and Spain in Fig. 5.10. We observe that since the early 1980s the inflation rates of Italy and Spain converged on the German one without, however, moving towards equality with it.

This situation led to a credibility problem. As the Italian and Spanish inflation rates remained above the German one for many years, the price *levels* of Italy and Spain tended to diverge continuously from the German one. Since there were no realignments after 1987 to compensate for these divergent price trends, a continuous loss of competitiveness of the Italian and the Spanish industries occurred.

We present the evidence for Italy and Spain in Fig. 5.11, which shows the Italian and Spanish price indices relative to the German one (expressed in

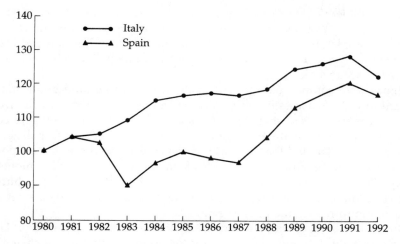

Fig. 5.11. Price index in Italy and Spain relative to Germany (expressed in common currency)

Source: EC Commission (1993).

common currency). It can be seen that during the 1980s Italian prices (after correction for the devaluations of the lira) increased by close to 30% relative to German prices. A somewhat smaller increase is noted in the Spanish case. This divergent trend in price levels jeopardized the competitiveness of the Italian and the Spanish industries. In the end this became unsustainable.

4.2. The sterling and French franc crises

The problems that arose with the pound sterling and the French franc were different in nature. The most striking feature of these crises is that there was very little evidence of divergent trends in prices and competitiveness of France or the UK relative to Germany. In fact, when comparing 'fundamental' economic variables of France and the UK with those of Germany (the current accounts of the balances of payment, for example), the former two countries scored as well as Germany. What then triggered the speculative crises?

Some continental European observers and politicians have claimed that the speculation against the pound sterling and especially against the French franc was irrational and driven by an 'Anglo-Saxon plot' against the process of monetary unification in Europe. Such explanations based on irrational motives cannot easily be disproved. One should be suspicious about these explanations, however. In general speculators want to make money, and do not care about the colour of the money they expect to earn.

A better explanation is available. It is based on what we have called the liquidity problem of fixed exchange rate regimes. At the start of the 1990s, and especially after 1992, Europe was hit by a severe recession. Very quickly this created a conflict, between Germany on the one hand, and Britain and France on the other hand, about the appropriate interest rate policy to be followed in the system as a whole.

Things were complicated by the fact that the German unification of 1990 had led to large increases in government spending in Germany which created inflationary pressures in that country. As a result, the Bundesbank gave complete priority to combating inflation by a restrictive monetary policy. The recession in the UK and France, however, demanded a looser monetary policy. A conflict arose because the Bundesbank maintained a restrictive monetary policy. The UK and France increasingly felt that this policy stance was hurting their economies, and pressured the German authorities to relax their monetary policies and to reduce the interest rates.

This policy conflict did not remain unnoticed by the speculators. These realized that the UK and French authorities were tempted to cut their links with the mark so as to be able to follow more expansionary monetary policies. Influential economists in fact openly urged the authorities to do exactly that. Thus, speculators had good reasons to start speculating against the pound sterling and the French franc. This happened first in September 1992 and led to the withdrawal of the pound sterling from the exchange rate mechanism of the EMS. From then on, as the speculators had expected, the UK authorities engaged in a policy of monetary expansion, and the pound sterling depreciated sharply in the foreign exchange markets.

One year later a new speculative crisis erupted involving mainly the French franc (but also the peseta, the Belgian franc, and the Danish kroner). The underlying reason was the same as in September 1992. The intensity of the recession in France during 1993 and the increse in unemployment were so great that many observers were convinced that the French government would have to stimulate the economy by lowering French interest rates. All this led to speculation of such a magnitude that the EC ministers of finance decided on 2 August to change the rules of the game. The margins of fluctuations were increased to +15% and −15%. This implied that the EMS currencies would be able to fluctuate in a band of 30%, transforming the system into a quasi-floating exchange rate regime. Although in a legal sense the EMS remained in existence, for all practical purposes the system ceased to exist.

4.3. The self-fulfilling nature of speculation

One of the interesting features of the speculative crises of 1992–3 is their self-fulfilling character. Speculators observe that the authorities have an incentive to change their policies. They also know that this can be done only by dropping out of the EMS. They then expect that this will happen and they start a speculation. In so doing, they force the authorities to drop out of the system.

This feature has been analysed in theoretical models, and should not have come as a surprise.[7] It has, however, led many people to think that the speculators are to blame for the collapse of the EMS. For it appears that, without objective reasons, they forced the authorities to drop out of the system. The next step to some sinister plot by 'Anglo-Saxon' speculators is a short one.

The self-fulfilling nature of speculation has also led some economists to propose the reintroduction of capital controls, thereby reducing the amount of funds that can be mobilized by speculators.[8] It is doubtful that capital controls would be able to sustain a fixed exchange rate, or that it would have prevented the disintegration of the EMS. As was argued earlier, a fundamental reason for the emergence of problems was the fact that the EMS evolved into a system of rigidly fixed exchange rates after 1987. The absence of capital controls certainly affected the timing and the dynamics of the disintegration of the EMS, but it did not fundamentally alter its instability, which resulted from the credibility and the liquidity problems of rigidly fixed exchange rate systems. After all, the Bretton Woods system collapsed for essentially the same reason, despite the fact that capital controls existed at the time of its collapse.

4.4. The role of Germany

It is sometimes said that one of the reasons for the collapse of the EMS is the fact that speculators can mobilize a vast amount of liquid funds, whereas the monetary authorities have only a limited amount of international reserves at their disposal. In this view, the monetary authorities are at the mercy of speculators, who possess superior weaponry. No wonder that the monetary authorities fought a losing battle.

This view is wrong. Against the large amount of speculative money there is an equally large amount of money the authorities can throw into the battle. In order to see this, it is useful to return to the two-country money market model of Section 2 (Fig. 5.3). When speculators start selling French francs, they buy German marks. The latter are sold by the Banque de France. As long as the Bundesbank is willing to supply marks to the Banque de France, there is no limit to the size of the interventions by the Banque de France. This has a simple reason. The Bundesbank creates the marks that it supplies to the Banque de France. In principle, it can create an unlimited amount of marks. Therefore, the monetary authorities can always win the battle against the speculators.

Why then did they lose? The answer is that the German authorities refused to continue supplying the French authorities with German marks. This refusal came about because the unlimited supply of German marks would have led to large increases in the German money stock. In other words, when

---

[7] See e.g. Obstfeld (1986).
[8] An example of such a proposal is Eichengreen and Wyplosz (1993).

the Bundesbank was supplying marks to the Banque de France, these marks increased the German money stock. As pointed out earlier, the Bundesbank was routinely sterilizing these increases of the supply of marks by reverse operations in the domestic money markets. The size of the interventions was, however, so high that the Bundesbank was not capable of completely sterilizing them. Thus, a massive support of the French franc would have forced the Bundesbank to relax its monetary policy stance. This the Bundesbank was unwilling to do.

Thus, the reason why the speculators won the battle was not that they had more weapons at their disposal than the monetary authorities. Rather, the reason is to be found in the fact that the monetary authorities, and in particular Germany, refused to throw all their weapons into the battle. This refusal was motivated by the fact that the unconditional defence of the French franc would have forced the German monetary authorities to change their policy stance. Therefore, as noted earlier, the fundamental reason for the disintegration of the EMS can be said to be the conflict about the appropriate monetary policy in the system and the refusal of Germany to follow a more expansionary monetary policy.

5. Credibility of the EMS: A formal analysis

The trends in the credibility of the EMS can be analysed in a formal way. In a number of papers Svensson has proposed several tests to find out whether the existing EMS parities were considered credible by foreign exchange market participants (see e.g. Svensson (1992)). The simplest test can be constructed as follows.[9] In efficient markets the forward exchange rate quoted today reflects the expectations that prevail about the future spot rate. Thus, if we take the one-year forward rate, for example, we have information about the expectations market participants have concerning the spot exchange rate in one year. Note that this test makes abstraction from the existence of risk premia. Some of the issues relating to risk premia are discussed in Box 7.

We can now find out whether market participants expect a realignment in the following way. The band of fluctuation defines the limits within which the exchange rates must remain if no realignment occurs. Thus, if the forward rate exceeds the upper limit, this is an indication that market participants expect the future spot rate (say, in one year) to be above the upper limit. In other words, they expect a revaluation of the foreign currency (a devaluation of the domestic currency). The opposite occurs if the forward rate is below the lower limit of the band of fluctuation.

In Fig. 5.12 we show the one-year and the five-year forward rates of the

---

[9] There are more complicated tests also. These give broadly the same results as the simple one explained here. See Rose and Svensson (1993).

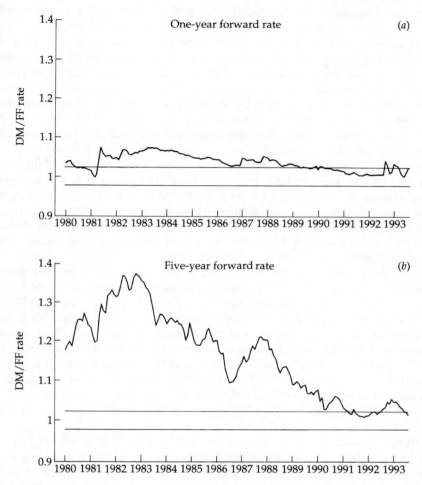

Fig. 5.12. Implicit forward German mark/French franc rates

Source: IMF, *International Financial Statistics*.

German mark relative to the French franc, together with the upper and lower limits as defined by the permissible band of fluctuation.[10] The spot exchange rates have been normalized to 1, so that we measure the forward rate as a percentage deviation from the spot rate. For example, a number of 1.1 means that the forward rate is 10% above the spot rate. Note also that we do not show the realignments. The latter imply that the band of fluctuation shifts up (or down).

---

[10] These forward rates are not quoted rates. They were computed by using the interest parity condition. In Box 7 we explain how this can be done.

Several observations can be made from Fig. 5.12. First, the five-year forward rate exceeds the upper limit much more than the one-year forward rate. This has to do with the fact that when agents expect a devaluation of, say, 5% per year, this cumulates to (more than) 25% over a five-year period. As a result, the five-year forward rate will be (more than) 25% above the spot rate. The one-year forward rate will only be 5% above the spot rate. Note that these expectations also show up in interest differentials. In Box 7 we make the link between the forward rates and interest differentials using the interest parity theory, and we show that the credibility tests can also be performed using these interest differentials.

Second, the one-year forward rate is more often within the band of fluctuation than the five-year forward rate. Except for a brief period in 1991, the latter was always outside the band. This suggests that whereas economic agents were quite often confident that a devaluation would not occur *during the next year,* they almost always perceived a devaluation risk over a five-year period. Thus, one can conclude from this that the permanent fixity of the mark/franc rate was almost never credible during the EMS period.

Third, the credibility of the fixity of the mark/franc rate improved continuously during the period. As mentioned earlier, during the early 1990s the fixity of the mark/franc rate came close to full credibility (especially during 1991).

For most other EMS currencies we observe similar phenomena. We show the evidence in Fig. 5.13 for the lira and the Belgian franc. Note that in the Italian case the margins of fluctuation were narrowed from +6% and −6% to +2.25% and −2.25% in 1990. We observe that the five-year forward rates of these two currencies stayed outside the credibility limits all the time, suggesting that economic agents never believed that these exchange rates would remain permanently fixed. We also observe a significant decline of the devaluation risk over time.

The mark/guilder rate is the only exchange rate in the EMS which appears to have been credibly fixed (in the sense that the five-year forward rate remained within its credibility limits). This is seen in Fig. 5.13, which shows that, at least after the middle of the 1980s, the five-year forward rate stayed within its credibility limits during most of the time.

The empirical evidence of the trends in credibility of the EMS raises a number of questions and puzzles. A first one is the following. During the first part of the 1980s we observe large deviations of the forward rate from the credibility limits, suggesting expectations of large realignments. In the early 1990s these deviations had declined significantly, suggesting much smaller expectations of devaluations. Why is it that the large expected devaluations of the early 1980s did not lead to a collapse of the EMS, whereas much smaller ones triggered the disintegration of the system? In order to answer this question we have to return to the discussion of the change in the nature of the EMS after 1987. Prior to that date there was an understanding in the market that small but frequent realignments were a routine matter. As a result, expectations of devaluation of, say, 5% per year could easily be accommodated

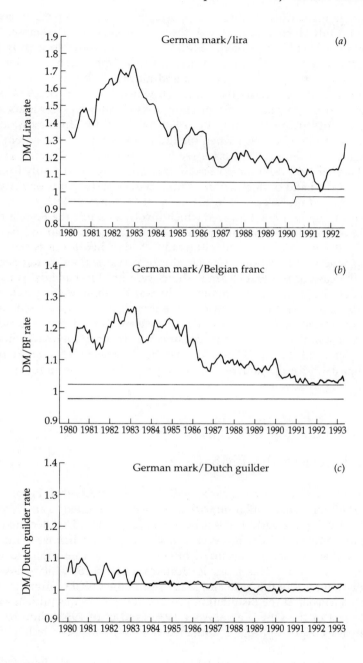

Fig. 5.13. Implicit forward five-year rates

Source: IMF, *International Financial Statistics.*

without triggering unsustainable speculative crises (see the discussion in section 3.1). Only during 1982–3 was the system close to a collapse. After 1987 it became the objective of the EMS countries to move to a truly fixed exchange rate system. As discussed earlier, this made the system too rigid and much more prone to speculative attacks, and ultimately led to its demise.

A second question concerns the degree to which the crisis of 1992–3 was anticipated. In order to analyse this question, we show the five-year mark/franc and mark/lira forward rates during the 1990s in Fig. 5.14. (In the case of the lira, we stop in September 1992 when the lira dropped out of the system.) We observe that the September 1992 crisis seems to have been anticipated. During the second half of 1991 both forward rates start deviating significantly from the upper limit of the band of fluctuations. Thus, market participants perceived an increasing devaluation risk prior to the crisis of September 1992.

The crisis of August 1993, however, which involved mainly the French franc, seems to have been largely unanticipated. This can be seen from the fact that during 1993 the five-year forward rate gradually returned towards the upper limit of the band of fluctuation. How can this be explained? As we have argued earlier, in the case of the French franc there were no 'fundamental' variables indicating that the franc was in trouble, in the way the lira and the peseta were. However, the source of the problem with the French franc must be seen in the policy conflicts that arose as a result of the deepening of the recession in Europe. These conflicts, and the ensuing uncertainty, only appeared when the recession intensified during 1993. This may be the reason why the exchange crisis of August 1993 erupted in an abrupt way and does not seem to have been anticipated long in advance.

6. Disinflation in the EMS

Disinflation in the EMS since the early 1980s has been substantial. From a peak of 11% in 1980, the rate of inflation within the system declined to an average of 2% in 1988. During the early 1990s it hovered around 3%. This decline of the inflation rate within the EMS, however, is not exceptional when we compare it with the disinflation that has occurred in the rest of the industrialized world during the same period. This is made clear from Fig. 5.15, which shows the average inflation rates in the EMS and in the rest of the OECD countries. From this figure we learn that the disinflation process in the EMS was initially slower than in the rest of the OECD area. After 1986 (and until 1992), however, the EMS countries caught up, and maintained a somewhat lower inflation rate than these other countries.

The low inflation rates achieved within the EMS since 1986 have led to a widespread belief that there is a disciplining feature in the EMS arrangement that facilitates disinflation. This view was discussed in previous sections, and

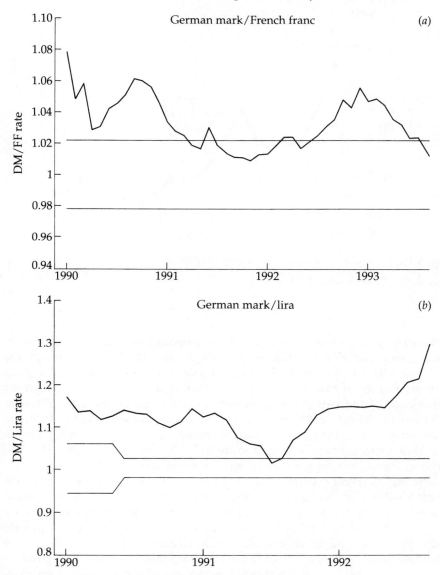

Fig. 5.14. Implicit forward five-year rates during the 1990s

Source: IMF, *International Financial Statistics.*

has been given a more formal content by Giavazzi and Pagano (1988), Mélitz (1988), and Giavazzi and Giovannini (1989).[11]

---

[11] For sceptical notes, see Collins (1988), Fratianni (1988), and Fratianni and von Hagen (1990).

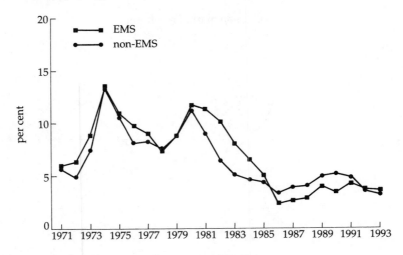

Fig. 5.15. Inflation (CPI) in EMS and non-EMS (OECD)

Source: OECD, *Main Economic Indicators.*

Policies of disinflation usually lead to temporary increases in unemployment. This was also the case in the 1980s. In Fig. 5.16 we present the unemployment trends in the EMS and in the non-EMS countries. The most striking feature of this figure is the fact that since 1983 the unemployment experience of the two groups of countries has been dramatically different. Whereas in the non-EMS OECD countries unemployment started declining substantially from 1983, it continued to increase in the EMS. During 1984–8 the EMS unemployment stabilized at around 10%. Only during 1989–90 did it decline. From 1991 on it increased again substantially.

The very different experience of EMS and non-EMS OECD countries is also made vivid in Figs. 5.17 and 5.18, which present the inflation-unemployment trade-offs in the two groups of countries. In both groups of countries the oil shocks of 1973–4 and 1979–80 worsen the trade-off. The disinflation process since the second oil shock coincides with a significantly higher increase in unemployment in the EMS than in the non-EMS OECD. In the latter group of countries we observe a strong inward movement of the inflation-unemployment trade-off since 1983. The EMS countries' trade-off remains stuck at a very unfavourable level from 1984 to 1988. Only during the boom period of 1989–91 does the trade-off improve somewhat.

These are certainly striking phenomena. How can they be explained? A first possible explanation relies on supply-side stories. It could be that labour

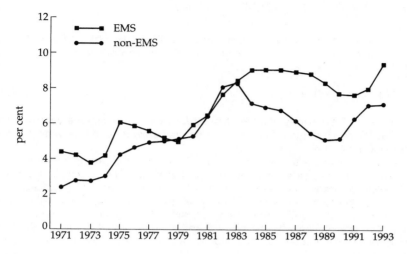

Fig. 5.16. Unemployment in EMS and non-EMS (OECD)
Source: OECD, *Main Economic Indicators.*

market rigidities have been more pronounced in the EMS countries than in the other OECD countries. As a result, the supply shocks of the early 1980s led to an increase in the natural level of unemployment, and prevented the inflation-unemployment trade-offs from moving inwards. Some evidence in support of this hypothesis is given in De Grauwe (1990).

This explanation, however, is not sufficient. It also appears that, at least initially, the EMS provided little of the disciplining effects. The reason was that there was little credibility in the fixity of the exchange rates (see the discussion of the previous section). Countries like France and Italy devalued regularly during the first half of the 1980s, and therefore imported little inflationary discipline from Germany. A change occurred around the middle of the 1980s. France and (to a lesser degree) Italy made it clear that they were committed to maintaining more rigid exchange rates with the low-inflation country in the system. This is also the time that the inflation record of these countries improved. However, we also observed that the unemployment rate remained stubbornly high. These changes certainly helped to improve the credibility of the fixed exchange rates within the system. They demonstrated to economic agents that the monetary authorities of these countries were giving an over-riding priority to keeping inflation low, even if this contributed to more unemployment.

Box 7. Interest parity and EMS credibility

The previous analysis of the credibility of the EMS can also be performed using interest rates. In order to do so, we introduce the interest parity theory, which we write as follows:

$$1 + r_F = \frac{F}{S}(1 + r_G) \tag{B7.1}$$

where r_F is the French (one-year) interest rate, r_G is the German (one-year) interest rate, F is the forward mark/franc rate for contracts expiring in one year, S is the spot mark/franc rate.

The left-hand side is the return of an investment of one franc in a French franc asset. The right-hand side is the return of an investment of one franc in a German mark asset, whereby the German mark is bought spot and sold forward. The latter operation (sometimes also called a swap) ensures that the investment in mark assets is really equivalent (in terms of risk) to the investment in franc assets. This is also the reason why the two returns must be equal. Equation (B7.1) is called the 'closed' interest parity condition and is a very basic relation in international finance.

Things are a little more complicated if the maturity of the investment is different from one year. We can generalize the interest parity equation as follows:

$$(1 + r_F)^{m/12} = \frac{F_m}{S}(1 + r_G)^{m/12} \tag{B7.2'}$$

where F_m is the forward rate of a contract expiring in m months. We now have added the exponent m/12 to the interest rates. An example makes clear why. Take a two-year forward contract (m = 24). The investment period is then two years, so that the interest rate return must be compounded over two years. The exponent is 2 (24/12). When the investment period is less than one year, the compounding factor is less than one. For example, if the investment period is three months, the compounding factor becomes 1/4 (3/12).

The next step in the analysis consists in postulating that the forward rate, F, reflects the prevailing expectations about the future spot exchange rate, i.e.

$$F_m = E(S_m) \tag{B7.3}$$

where $E(S_m)$ is the expectation today about the spot exchange rate in m months. Replacing F_m by $E(S_m)$ in equation (B7.1') yields the 'open' interest parity conditions.[12]

---

[12] The relation with the open interest parity formula used in section 2 can be seen as follows. In order to simplify the notation we use the one-year interest parity condition. Substitute (B7.3) into (B7.1):

$(1 + r_F) = (E(S)/S)(1 + r_G)$.

Subtract $(1 + r_G)$ from both sides and rearrange. This yields

$(r_F - r_G)/(1 + r_G) = (E(S) - S)/S$ or

$(r_F - r_G)/(1 + r_G) = \mu$.

Thus, the open interest parity formula used in Sect. 2 differs from the formula derived here by the term $(1 + r_G)$. If the interest rates are not too high, this is a number close to 1, so that the formula used in Sect. 2 is approximately correct.

Equation (B7.3) assumes absence of risk premia. The consensus today is that these risk premia are real. However, the empirical evidence indicates that these risk premia are extremely variable, and that they cannot be explained very well. In fact, it is fair to say that attempts to find systematic movements in these premia have failed. This is also the reason why we abstract from these risk premia here.

The credibility test referred to earlier consists in checking whether

$$S_L < F_m = E(S_m) < S_U \tag{B7.4}$$

where S_L is the lower limit of the permissible band of fluctuation and S_U is the upper limit of the permissible band of fluctuation. Note that in the credibility tests reported in Section 5, we computed F_m using equation B7.2.

It can now easily be seen that the same credibility test can be interpreted by focusing on the interest rates. The limits imposed on F_m defined by equation (B7.4) also place limits on r_F in equation (B7.2). Using (B7.4), these limits can now be written as

$$\left(\frac{S_L}{S}\right)^{12/m} (1 + r_G) - 1 < r_F < \left(\frac{S_U}{S}\right)^{12/m} (1 + r_G) - 1. \tag{B7.5}$$

If the French interest rate remains within the limits defined by (B7.5), we can say that market participants had confidence that the mark/franc rate would not be officially devalued or revalued in the future.

A failure of credibility to hold can now be interpreted in terms of interest rates. When market participants expect a devaluation of the franc, they want to be compensated by an excess return on investments in francs. This then shows up in French interest rates which exceed the upper limit defined in equation (B7.5). If confidence in the French franc increases, the excess return declines. This also has the effect of bringing the French interest rates progressively closer to the German interest rate. Thus, a narrowing of the interest differentials can be interpreted as an increase in the credibility of the fixed exchange rates.

Note that with an increase in the maturity of the investments the band within which the interest rate must stay becomes progressively narrower. The reason is that, when a depreciation of, say, 1% per year is expected, during ten years this will cumulate to an expected devaluation of 10%. The interest rate on a ten-year French bond will then be 1% above the German bond. Thus, a 1% interest differential on *ten-year* bonds indicates that the market expects a 10% devaluation of the franc to occur during the next ten years. The same 1% interest differential on a *one-year* bond implies that agents expect a depreciation of 1% of the franc next year. Since this keeps the franc within the band of fluctuation, this 1% interest differential does not mean that a realignment is expected next year.

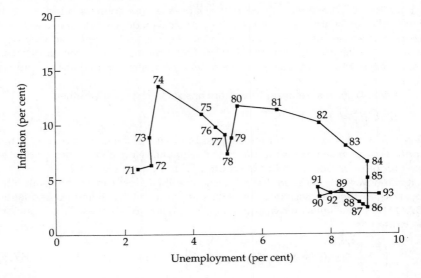

Fig. 5.17. Inflation-unemployment in EMS

Source: OECD, *Main Economic Indicators.*

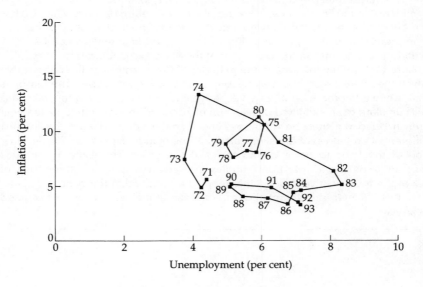

Fig. 5.18. Inflation-unemployment in non-EMS (OECD)

Source: OECD, *Main Economic Indicators.*

One can conclude that the evidence of strong disciplining effects in the EMS is weak. Only during the second half of the 1980s is there some evidence that the inflation discipline imposed by the system started to work. It is thus not clear whether the EMS reduced the cost of disinflation, compared to other monetary arrangements.

7. The EMS and the recession of 1992–3

As argued earlier, the disintegration of the EMS in 1993 had much to do with the unprecedented recession that occurred in Europe in 1992–3. This recession exacerbated the policy conflicts between the major members of the system, and contributed to the downfall of the system.

It can also be argued that, while the recession destroyed the EMS, the EMS tended to intensify the recession in Europe.[13] The two-country money market model presented in Section 2 allows us to understand why. The recession in the peripheral countries shifted these countries' money-demand functions to the left. In Germany there was also a recession. However, at the same time the increased government spending in East Germany had increased the rate of inflation and induced the Bundesbank to restrict the money supply. A net upward movement in the German interest rate ensued, which forced the authorities of the periphery to follow suit.

In Fig. 5.19 we present some data that illustrates the degree of monetary restriction applied by EMS countries during the recession of 1992–3. We selected the real short-term interest rate as an indicator of the degree of monetary restriction. It can be seen that the EMS countries kept their short-term interest rates at relatively high levels during their worst post-war recession.

We also contrast the behaviour of the EMS interest rates with those of the USA. It can be seen that the USA followed quite a different monetary policy during its recession (which started two years earlier than in the EMS countries). At the start of the US recession in 1989, the Federal Reserve lowered the short-term interest rate significantly. The EMS countries followed exactly the opposite monetary policy, thereby aggravating the recession. Only in 1993 was there some monetary relaxation. This still left real short-term interest rates at 4% or more.

---

[13] It can be argued that the EMS *per se* is not responsible for intensifying the recession, but that the lack of policy co-ordination is. However, one can still argue that the EMS may not have given the right incentives to co-operate.

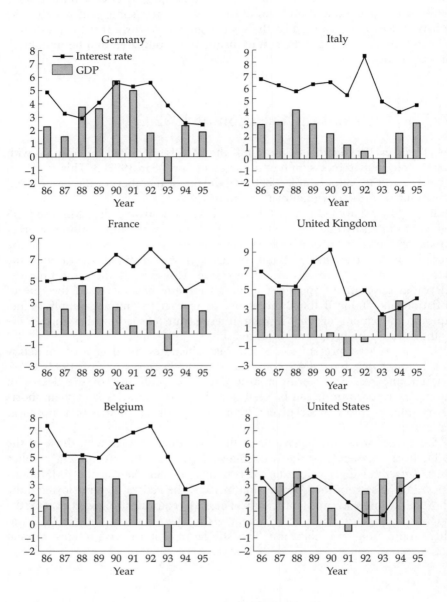

Fig. 5.19. Real short-term interest rate and real GDP (growth rate)

Source: EC Commission, *European Economy*.

It is useful to ask the question what would have happened if a European central bank had been in existence during the recession. Such a central bank would have cared, not about monetary conditions in Germany, but rather about the monetary conditions in the whole system. Thus, suppose that the European central bank would have followed a Friedman-type rule of targeting the European money stock. Such a European central bank would have targeted the *sum* of the money stocks of Germany and of all the other EMS countries. Thus, while there might have been reasons for restriction in Germany, there was no reason to follow restrictive monetary policies in the other EMS countries. Put differently, a monetarist policy rule applied by a European central bank during the recession would have allowed for a more expansionary monetary policy in Europe than the one that was applied in the EMS. It is in this sense that the EMS led to an excessively deflationary monetary policy and intensified the recession in Europe.

Box 8. Target-zone models and the European Monetary System

The exchange rate mechanism of the EMS operates with a band within which the exchange rate can move freely. Once the edges of the band are reached, however, the monetary authorities are committed to intervene so as to prevent the exchange rate from moving outside the band. This feature of the system had led theorists to speculate about the behaviour of the exchange rate in such a system. This has led to the so-called target-zone models poineered by Krugman (1989) and developed further by (among others) Svensson (1990) and Bertola and Svensson (1993).

The basic idea of these models is strikingly simple (although the mathematics is not). Suppose speculators are fully confident that the exchange rate will not go beyond the limits of the band because they are convinced that the monetary authorities will successfully defend these limits. Assume in addition that there are stochastic disturbances in the fundamental variables that determine the exchange rate. (These fundamentals could be the money stock, the price level, etc.) How is the relationship between these fundamentals and the exchange rate going to look? The answer is given in Fig. B8.1. On the vertical axis we show the exchange rate (S_t), on the horizontal axis the fundamental variable (f_t). The two horizontal lines (S_U and S_L) represent the limits between which the exchange rate will fluctuate. The fluctuations of the fundamental variable lead to fluctuations of the exchange rate. In a world of rational expectations, the exchange rate must lie on the S-shaped curve. That is, as the exchange rate comes close to, say, the upper limit of the band, speculators know that the authorities' commitment will prevent the exchange rate from moving beyond the limit. Therefore, the probability that the exchange rate increases in the future declines as we move closer to the limit, and the probability of a decline increases. Speculators will find it advantageous to sell foreign exchange, thereby pushing the exchange rate downwards. Speculation will be stabilizing, and the authorities will not have to intervene. This result only holds if the credibility of the band is very strong. If this is not the case, speculation will not be stabilizing. Speculators then will test the resolve of the authorities, forcing them to intervene in

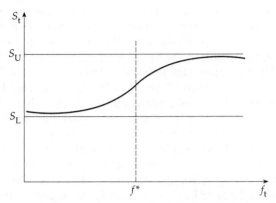

Fig. B8.1. The exchange rate and the fundamental variable in a target-zone model

the market. The model has been extended by Bertola and Svensson (1993) to this case where speculators expect a devaluation. They show that if the devaluation risk is strong enough, the exchange rate will not follow an S-shaped curve as in Fig. B8.1, but may actually cross the margins.

8. The EMS since 1993

By enlarging the normal band of fluctuation from $2 \times 2.25\%$ to $2 \times 15\%$ the EMS changed its nature in a very drastic way. As mentioned earlier, only one country, the Netherlands, maintained the previous narrow band. Some countries, like Belgium and Austria, kept the bands of fluctuation of their currencies with the DM within an informal narrow band. In general countries have not exploited the full flexibility of the wide band and have tried to steer the exchange rates within an informal and narrower band than the 30% one, without however committing themselves to defending the limits of these narrower bands.

The post-1993 developments within the EMS have been characterized by relative stability (despite the fact that the trends towards more capital mobility have continued). This relative stability has much to do with a feature we identified in Section 3. With a total band of fluctuation of 30% the 'one-way bet' feature of the EMS disappeared, so that the opportunities for large speculative gains of a speculative attack vanished almost completely. The reason is very simple. It has become extremely unlikely that in the event of a realignment, this will be more than 30%. The underlying differences in fundamentals among EMS countries (inflation rates, for example) have become quite small. As a result, large realignments are unlikely to happen in the foreseeable future. Thus, if speculators attack a currency, they know that the possible ensuing devaluation will remain small relative to the wide band of 30%. The effect of their attack would be to push up the foreign exchange rate within the band. If the realignment then happens, it is very unlikely that this will lead to a discrete jump in the exchange rate after the realignment since the new lower limit is most likely to fall below the previous upper limit. This is the situation we presented in Fig. 5.7.

We conclude that the existence of a wide band of fluctuation, together with the relative stability of underlying fundamental variables, has shielded the EMS since 1993 from the disruptive effects of speculative attacks.

9. Conclusion

In this chapter we have analysed the workings of the European Monetary System. We argued that after 1987, when the system evolved into a truly fixed exchange rate regime, it was unable to cope with the problems that have plagued every fixed exchange rate arrangement of the past. Its downfall therefore did not come as a surprise and had been predicted by many economists.

Although the disintegration of the EMS was, and could be, predicted, the particular way it happened, of course, could not be foreseen. In this chapter, great emphasis was put on the recession in Europe of 1992–3 as an explanation of the downfall of the system. This recession exacerbated the conflicts between the major EMS countries about the appropriate monetary policy response. The inability to resolve this policy conflict lies at the root of the loss of confidence of economic agents in the fixity of the exchange rates, and in the ensuing speculative crises. In addition, we argued that the EMS intensified the recession in Europe. A return to such an exchange rate arrangement would certainly not be desirable.

What then are the options for the future? There are basically two possibilities. One is a return to a more flexible exchange rate arrangement than the one that existed during 1987–92. In a sense this is the option that the European countries chose in 1993 (at least temporarily) by drastically enlarging the permissible band of fluctuation of the exchange rates. The other alternative consists in moving towards a full monetary union. In the next chapter we study that alternative, and we identify the problems and the obstacles on the road towards a full monetary union in Europe.

6

THE TRANSITION TO A MONETARY UNION

Introduction

In December 1991 the heads of state of the European Union signed an historic treaty in the Dutch city of Maastricht. The Maastricht Treaty went well beyond purely monetary affairs. Nevertheless it is best known for the blueprint it provided for progress towards monetary unification in Europe.

The Maastricht Treaty strategy for moving towards monetary union in Europe is based on two principles.[1] First, the transition towards monetary union in Europe is seen as a gradual one, extending over a period of many years. Second, entry into the union is made conditional on satisfying convergence criteria. In this chapter we analyse this Maastricht strategy.

It is important to be aware that the Maastricht strategy was not the only one available. In fact, throughout history monetary unification has quite often been organized in a very different way. Take as an example the German monetary unification, which happened on 1 July 1990. The characteristic feature of the German monetary union was its speed and the absence of any convergence requirement. The decision to go ahead with monetary union was taken at the end of 1989, and six months later the German monetary union was a reality. As will be seen in the next section it will take the European Union at least ten years to do the same. In addition, East Germany was allowed into the West German monetary area without any conditions attached. Surely, had Maastricht-type convergence requirements been imposed on East Germany, the German monetary union would not have occurred. All this shows that a monetary union *can* be established quickly and without prior conditions. It does not show, of course, that this is the desirable way to organize a monetary union in Europe. One of the questions we will analyse in this chapter is precisely whether the Maastricht approach was the right way to move towards monetary union in Europe.

---

[1] In this the Treaty was very much influenced by the Delors Committee Report, which was issued in 1989. See Committee on the Study of Economic and Monetary Union (1989).

1. The Maastricht Treaty

The approach set out in the Treaty is based on principles of gradualism and convergence. Let us analyse the details of this approach. The Treaty defines three stages in the process towards monetary union.

In the first stage (which had already started on 1 July 1990, prior to the signing of the Treaty), the EMS countries abolished all remaining capital controls. The degree of monetary co-operation among the EMS central banks was strengthened. During the first stage, which lasted until 31 December 1993, realignments remained possible.

The second stage started on 1 January 1994. A new institution, the European Monetary Institute (EMI), was created. It operates only during this second stage, and is in a sense the precursor of the European Central Bank (ECB). Its functions are limited, and are geared mainly towards strengthening monetary co-operation between national central banks.

At the start of the third and final stage the exchange rates between the national currencies will be irrevocably fixed. In addition, the European Central Bank will start its operations. The ECB will issue the European currency, which will become a currency in its own right. The transition to this final stage of monetary union, however, is made conditional on a number of 'convergence criteria'. A country can only join the union if:

(1) its inflation rate is not more than 1.5% higher than the average of the three lowest inflation rates in the EMS;

(2) its long-term interest rate is not more than 2% higher than the average observed in the three low-inflation countries;

(3) it has not experienced a devaluation during the two years preceding the entrance into the union;

(4) its government budget deficit is not higher than 3% of its GDP (if it is, it should be declining continuously and substantially and come close to the 3% norm, or alternatively, the deviation from the reference value (3%) 'should be exceptional and temporary and remain close to the reference value', art. 104c(a));

(5) its government debt should not exceed 60% of GDP (if it does it should 'diminish sufficiently and approach the reference value (60%) at a satisfactory pace', art. 104c(b)).

If a majority of EMS countries had satisfied these conditions, the third stage could already have started at the end of 1996. However, the ECOFIN decided in November 1995 that there was no majority to start the third stage at the end of 1996. The Treaty then stipulates that the third stage has to start at the latest on 1 January 1999, with those countries that satisfy the criteria. The UK has maintained the right to opt out, and Denmark to subject the agreement to a national referendum.

From the wording of the Treaty it is clear that the convergence criteria (especially the budgetary ones) are subject to interpretation. For example, the condition that the government debt ratio should 'decline sufficiently' and 'approach the reference value of 60% at a satisfactory pace' can lead to disagreement and conflict. This will have to be resolved. In this connection, the Treaty stipulates the following decision procedure. The European Commission and the European Monetary Institute shall make a report determining the progress each country has made towards convergence and will propose whether the country should be accepted into the EMU. The final decision, however, will rest with the Council of Ministers, who will decide about membership for each country using the qualified majority rule.

At the Summit Meeting of the heads of state in Madrid, December 1995, additional agreements were made concerning the nature of the third stage. First, it was decided to call the new currency the euro. Second, the third stage was itself divided into three substages as follows:

● From 1 January 1999 until 31 December 2001, the national currencies will continue to be in circulation alongside the euro, albeit at irrevocably fixed exchange rates. However, commercial banks will use the euro for all their interbank dealings. Private individuals will have the choice of using their national currency or opening an account in euros. (Note that during this period the euro will not exist in the form of banknotes and coins). In addition, all transactions between the European Central Bank and the commercial banks will be in euros. Finally, *new* issues of government bonds will be made in euros and not in national currencies.

● During the period 1 January to 1 July 2002 the euro will replace the national currencies, which will lose their legal-tender status. Thus, during this period a monetary reform will be organized.

● From 1 July 2002 on, a true monetary union will come into existence in which the euro will be the single currency managed by one central bank, the European Central Bank (ECB).

It should be noted that the national central banks will not disappear after 2002. They will be part of what is called the European System of Central Banks (ESCB). These national banks, however, will not make decisions about monetary and exchange rate policies any more. They will be there to implement the decisions taken by the ECB. In this respect the ESCB will resemble the US Federal Reserve System. It should be noted, however, that national central banks will maintain their decision-making powers in the important field of banking supervision.[2] (We will return to some issues relating to the operation of the ESCB in Chapter 7).

Finally, the Madrid Summit Meeting also determined that decisions about membership in EMU should be taken in early 1998 based on the macroeconomic performances of the individual countries during the year 1997. In other words, countries will be judged in early 1998 as to whether they achieved economic convergence in 1997. If they do they will be allowed into EMU from 1 January 1999.

---

[2] In some countries the responsibility for banking supervision is not vested in the national central bank but in a separate agency. In these countries the national bank will have few responsibilities left over.

One potential problem, to which we will return, is that while the decision about membership will be taken in early 1998, the decision concerning the level of exchange rates (the conversion rates) that will be used to lock currencies irrevocably to each other will be taken on 1 January 1999. This creates a period of uncertainty about these conversion rates which may invite speculative pressures. We will also have to return to that issue. The only precise statement to be found in the Treaty about how these conversion rates will be fixed is that this decision shall be taken unanimously by all countries accepted into EMU.

2. Why convergence requirements?

We noted earlier that past transitions to monetary unions were usually organized in a different way than in the Maastricht Treaty, i.e. once the decision was taken to have a monetary union, this was done quickly without any of the Maastricht-type convergence requirements being imposed on the prospective members. What is more, the theory of optimum currency areas, which we discussed in previous chapters, is silent about Maastricht-type convergence criteria. Instead the OCA theory stresses the need to have labour market flexibility and labour mobility as important requirements for a successful monetary union. According to this theory, if these conditions are satisfied, there is no need to wait more than ten years to do it. Why then have the designers of the Treaty stressed so much *macro*-economic convergence (inflation, interest rates, budgetary policies) prior to the start of EMU while the theory stresses *micro*-economic conditions for a successful monetary union?

2.1. Inflation convergence

The answer has to do with the fear that the future monetary union would have an inflationary bias. In order to understand this concern it is useful to go back to the Barro–Gordon model which we developed in Chapter 3, and which we now represent in Fig. 6.1. We assume two countries, called Germany and Italy. The two countries are assumed to be identical except for the preferences of the authorities. (We do not really need this assumption. We do this only to be able to put both countries in the same figure.) The German authorities give a high weight to reducing inflation, the Italian authorities a low weight. This is shown by flat indifference curves for the German authorities and steep ones for the Italian authorities. The natural unemployment rate, u_N, is the same in the two countries, and so is the target unemployment rate of the authorities, u^*. Inflation equilibrium is achieved at E_G in Germany and E_I in Italy. Thus, inflation is on average higher in Italy than in Germany without any gain in unemployment for Italy.

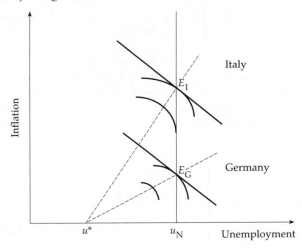

Fig. 6.1. The inflation bias in a monetary union

A monetary union between the two countries implies that a common central bank takes over. Two propositions can now easily be established. First, the low-inflation country (Germany) always reduces its welfare by forming a monetary union with the high-inflation country. This is so because the union's central bank is likely to reflect the average preferences of the participating countries. As a result, the union inflation rate increases and will be located between E_G and E_I. (There are of course other sources of gains of a monetary union, e.g. lower transaction costs, lower risk, etc. which we discussed in Chapter 4, and which are outside the model of Fig. 6.1. These efficiency gains must then be compared with the welfare losses resulting from higher inflation. If the latter exceed the former, Germany will not want to join in a monetary union with Italy.)

The second proposition follows from the first one: since the low-inflation country, Germany, loses when it joins the union with Italy, it will not want to do so unless it can impose conditions. It follows from the analysis of Fig. 6.1 that this condition must be that the union's central bank should have the same preferences as the German central bank. This can be achieved in two ways. One is that Germany will insist that the future European central bank should be a close copy of the Bundesbank. What this means will be analysed in the next chapter.

This condition, however, may not be sufficient from the point of view of Germany, because the European Central Bank will be composed of representatives of the participating countries. Even if the ECB is made a close copy of the Bundesbank, these representatives may still have different inflation preferences. Majority voting in the Board may then put the German representative in a minority position, so that the equilibrium inflation rate in the union would exceed the German one. In order to avoid this outcome Germany will want to control entry into the union, so that only those countries with the same preferences join the union (see Morales and Padilla (1994)).

The Maastricht entry conditions can now be interpreted in this perspective. Before the union starts, the candidate member countries are asked to provide evidence that they care about a low inflation rate in the same way as Germany does. This they do, by bringing down their inflation rate to the German level. During this disinflationary process, a temporary increase in the unemployment rate will be inevitable (a movement along the short-term Phillips curve). This self-imposed suffering is added evidence for Germany that countries like Italy are serious about fighting inflation. Once the proof is given, these countries can be let in safely.

This interpretation of the Maastricht convergence requirements has now become the conventional one. It certainly sounds plausible. We will analyse it critically in Section 3. At this stage it is already important to stress that this argument for prior reduction of inflation by Italy is not really a condition necessary for forming a monetary union. It is a condition of forming a monetary union with a low inflation rate. There is no reason why a monetary union between Germany and Italy could not work with an inflation rate that would be the average of the German and the Italian inflation rates observed before the monetary union. To put it in more concrete terms. Since 1960 the yearly inflation rates in Germany and Italy have been 3.4% and 8.9% on average. A European monetary union in which the rate of inflation would be an average of these two numbers, i.e. 6.1%, could certainly function. The problem is that the low-inflation country does not want this outcome.[3]

2.2. Budgetary convergence

Can the budgetary convergence requirements (3% norm for the budget deficit and 60% norm for the government debt) be rationalized in a way similar to the inflation convergence requirement? The answer is positive. Let us take again the case of Italy and Germany. Italy has a high debt-to-GDP ratio (more than 100% during the 1990s). A high government debt creates incentives for the Italian government to engineer a surprise inflation. The reason is that some of the Italian government bonds are long-term. The interest rate on these bonds was fixed in a previous period based on the then prevailing expectations of inflation. If the government now creates an unexpectedly higher inflation the real value of these bonds will be eroded and the bondholders will obtain insufficient compensation because the interest rate on their bonds does not

---

[3] We are not arguing that an average yearly rate of inflation of 6.1% is as good as one of 3.4%. There are many reasons to believe that the latter is better than the former. The point is that EMU would not be less of a monetary union when inflation is 6.1% than when it is 3.4%.

reflect this inflation upsurge. Bondholders lose. The Italian government gains. (Obviously, if the bondholders are rational, they will not be willing to invest in Italian bonds any more unless they obtain an extra risk premium on these bonds. Thus, the systematic use of surprise inflation by the Italian authorities may become quite costly in the long run. Rational governments, therefore, will not systematically produce surprise inflation. The problem here is that the political system may create a very short-term outlook for politicians, who will continue to be tempted to create inflation surprises.)

From the preceding analysis it follows that a monetary union between Germany and Italy creates a problem for Germany. In the union, the German authorities will be confronted with a partner who will have a tendency to push for more inflation. This may happen even if the German and Italian authorities have the same preferences regarding inflation. As long as Italy has a higher debt-to-GDP ratio it will have an incentive to create surprise inflation. As a result, Germany stands to lose and will insist that Italy's debt-to-GDP ratio be reduced prior to entry into the monetary union. In order to achieve this, Italy must reduce its government budget deficit. Once this is achieved the incentives for Italy to produce surprise inflation disappear, and Italy can safely be allowed into the union.[4]

Note again that the argument for debt and deficit reduction prior to entry into EMU is made not because countries with high debt and deficits cannot form a monetary union, but because allowing these countries into the union increases the risk of more inflation in the future EMU.

Other arguments have been developed to justify deficit and debt reductions as conditions for entry into the union. One is that the authorities with a large debt face a higher default risk. If they are allowed into the union, this will increase the pressure for a bailout in the event of a default crisis. The fear that this may happen also explains the *no-bailout* clause which was incorporated into the Maastricht Treaty, i.e. the clause that says that neither national governments nor the European Central Bank can be forced to bail out other member countries. In Chapter 8, where we discuss fiscal policies in monetary unions, we will return to this issue.

Whereas serious arguments can be found to justify the requirements that countries should reduce their government debts and deficits prior to entry (we will criticize these arguments, however), the numerical precision with which these requirements have been formulated is much more difficult to rationalize. This has led many economists to criticize the 3% and the 60% norms as arbitrary, or worse as some form of voodoo economics.[5] But let us at least

---

[4] It can be argued that if the Italian authorities reduce the maturity of their debt, the incentives to create surprise inflation are reduced. Thus as a substitute for a debt reduction, one could ask the Italian authorities to reduce the maturity of their debt prior to entry into the EMU. The problem with this is that when the maturity of the debt shortens, it becomes more vulnerable to changes in the interest rate, which may lead to liquidity crises as the authorities find it difficult to roll over their debt.

[5] See Buiter *et al.* (1993) and Wickens (1993) among others.

find some rational grounds for imposing the budgetary numbers of 3% and 60%.

The 3% and 60% budgetary norms seem to have been derived from the well-known formula determining the budget deficit needed to stabilize the government debt:[6]

$$d = gb \tag{6.1}$$

where b is the (steady state) level at which the government debt is to be stabilized (in per cent of GDP), g is the growth rate of nominal GDP, and d is the government budget deficit (in per cent of GDP).

The formula shows that in order to stabilize the government debt at 60% of GDP the budget deficit must be brought to 3% of GDP if and only if the nominal growth rate of GDP is 5% ($0.03 = 0.05 \times 0.6$).

The rule is quite arbitrary on two counts. First, it is unclear why the debt should be stabilized at 60%. Other numbers, e.g. 70% or 50%, would do as well. In that case, the deficit to be aimed at should also be different, i.e. 3.5% and 2.5% respectively. The only reason why 60% seems to have been chosen at Maastricht was that at that time this was the average debt-to-GDP ratio in the European Union. Second, the rule is conditioned on the future nominal growth rate of GDP. In order to illustrate the arbitrary nature of the Maastricht debt and deficit rules the nominal growth rates of the sixties, the seventies, and the eighties will be selected. What would have happened to the government debt if EU countries had applied the 3% deficit rule imposed by the Maastricht Treaty? The answer is given in Table 6.1.

We observe that applying the Maastricht 3% deficit rule would have led to very different steady-state government debt levels during the sixties, the seventies, and the eighties. In none of these decades would the application of the 3% deficit rule in the EC-12 have led to the 60% steady-state government debt. We have also listed the individual countries' steady-state debt levels. We find that only in the case of Germany during the eighties is the steady-state debt level implied by the Maastricht Treaty consistent with a 3% deficit rule. In all the other countries the steady-state government debt levels are much lower

---

[6] This is confirmed in Bini-Smaghi *et al.* (1993). The formula is derived as follows. The budget deficit (D) is financed by issuing new debt:

$$\dot{B} = D$$

(where B is the government debt and a dot above a variable signifies a rate of increase per unit of time). By definition one can write

$$\dot{B} = \dot{b}Y + b\dot{Y}$$

(where $b = B/Y$ and Y is GDP). Combining the two expressions yields

$$D = \dot{b}Y + b\dot{Y} \text{ or } D/Y = \dot{b} + b\dot{Y}/Y$$

which is rewritten as follows

$$\dot{b} = d - gb \text{ (where } d = D/Y, g = \dot{Y}/Y).$$

In the steady state $\dot{b} = 0$ which implies that $d = gb$.

Table 6.1. Steady-state government debt with 3% deficit (in per cent of GDP)

	1960s	1970s	1980s
EC-12	32	21	33
Germany	36	37	59
France	29	22	34
Italy	29	16	23
UK	42	19	33

Source: EC Commission (1993).

than 60%. This means that for these countries (with a higher growth rate of nominal GDP) the 3% deficit norm would have been more constraining than for Germany, forcing them to reduce the government debt to lower levels. It is interesting to note that the Maastricht budgetary rules seem to have been tailor-made for Germany (during the eighties).

The computations of the steady-state levels of debt in Table 6.1 are, of course, heavily influenced by the existence of different levels of inflation during the 1960s, 1970s and 1980s. One may argue that in a future monetary union with low inflation, these differences will tend to become less important. But what if inflation does not turn out to be low? After all, we do not know what the inflation performance in a future monetary union will be. It is, therefore, quite problematic to define in a constitution what these debt levels should be, as they are dependent on the rates of inflation that will prevail in the future.

We have focused on the inflation and budgetary convergence requirements in the preceding paragraphs. Let us briefly discuss the rationale of the other convergence rules, i.e. the no-devaluation rule and the interest rate convergence requirement.

2.3. Exchange rate convergence (no-devaluation requirement)

The main motivation for requiring countries not to have devalued during the two years prior to their entry into the EMU is straightforward. It prevents countries from manipulating their exchange rates so as to force entry at a more favourable exchange rate (a depreciated one, which would increase their competitive position). The stringency of this requirement, however, has been reduced considerably since the Treaty of Maastricht was signed. This has to do with the peculiar way the no-devaluation condition is formulated in the Treaty. According to the Treaty, countries should maintain their exchange rates within the 'normal' band of fluctuation (without changing that band) during the two years preceding their entry into the EMU. At the moment of the signing of the Treaty, the normal band was 2 × 2.25%. Since August 1993, the 'normal' band within the EMS has been 2 × 15%, a considerably larger band of fluctuation. There is still a legal dispute over whether the term 'normal' refers to what was normal at the time of the signing of the Treaty or whether it refers to the present situation. The consensus today seems to be that the latter interpretation should be followed. As a result, the exchange rate requirement has become quite soft. (It will be argued in Section 7 that this softer exchange rate rule can make the transition to monetary union easier than the more stringent rule envisaged at Maastricht.)

2.4. Interest rate convergence

We come finally to the interest rate convergence requirement. The justification of this rule is that excessively large differences in the interest rates prior to entry can lead to large capital gains and losses at the moment the EMU takes effect. Suppose, for example, that the long-term bond rate in Germany is 5% whereas in France it is 7% just prior to the start of EMU. When EMU starts the exchange rate between the DM and the FF will be irrevocably fixed. As a result, it will suddenly be quite attractive for bondholders to arbitrage, i.e. to sell low-yield DM bonds and to buy high-yield FF bonds. Since the exchange rate is irrevocably fixed there is no exchange risk involved in this arbitrage. As a result, it will go on until the return on DM and FF bonds is equalized. This will lead to a drop in the price of DM bonds and an increase in the price of FF bonds, until the yields are equal. Thus, economic agents (mainly German financial institutions) holding DM bonds prior to the start of EMU will make capital losses, and economic agents holding FF bonds (mainly French financial institutions) will make capital gains. These capital gains and losses increase with the size of the interest differential. When they are large enough they could create large disturbances in national capital markets. In order to limit these disturbances the interest differential should be reduced prior to the start of EMU.

The peculiarity of this rule is its self-fulfilling nature. The rule says that the long-term government bond rate of a prospective member should not exceed the interest rate level (+2%) of the three countries with the lowest rates of inflation. Consider now a country that is strongly expected to be a member of EMU after 1999. It can easily be seen that the long-term bond rate will be automatically equalized prior to the start of EMU. In this case it is really redundant. Note also that the capital gains and losses (which are inevitable) will have been borne long before the start of the union. At the start of the union, these capital gains and losses will be very small. All this will happen automatically even if the condition is not formulated explicitly. Consider now another country that is not expected to be in the EMU. In that case the interest differential is likely to remain high, thereby leading to non-fulfilment of a Maastricht convergence criterion. The latter then validates the expectations that the country will not be accepted. We have a self-fulfilling expectation: because agents expect a country not to become a member of EMU, it cannot become a member. Conversely, it suffices that agents change their expectations for the interest convergence criterion suddenly to cease to be an obstacle. Put differently, the expectation that there is an obstacle creates an obstacle; the expectation that there is no obstacle eliminates the obstacle. This is certainly a strange criterion to select countries for EMU.

3. Problems with the Maastricht strategy

In our analysis of the problems of the Maastricht strategy we will concentrate mainly on the inflation and budgetary convergence requirements. However, as will become clear, these convergence criteria also interact with the exchange rate and the interest rate criteria. We will criticize the Maastricht convergence requirements at three different levels. First, we analyse the criteria from the Europe-wide perspective. We ask the question whether the Maastricht strategy may not have imposed undue deflationary forces on the European Union, which in turn have made the budgetary norms more difficult to achieve. Second, we analyse the risk produced by the Maastricht convergence criteria for the unity of the European Union, and third, we study whether these criteria really achieve what they were intended to, i.e. to guarantee that the future monetary union produces low inflation.

3.1. The Maastricht strategy and the European business cycle

A striking fact is that during the 1990s economic growth in the group of countries that have declared their intention to follow the Maastricht transition strategy has been low compared to the previous decade and compared to the industrial countries not involved in the Maastricht strategy. In Table 6.2 we show the evidence. We contrast the average growth rate in the EU and the US. Whereas during the 1980s the growth rate of GDP was approximately the same in the EU and the US, this was not the case any more during the 1990s when the EU growth rate dropped significantly below the US level.

All this, of course, could be due to coincidence.[7] There are, however, reasons to believe that the Maastricht strategy is at least partially responsible for the lacklustre economic growth observed during the 1990s. The main problem of the Maastricht convergence criteria is that they imposed a policy mix of budgetary *and* monetary restriction. Countries asked to reduce their government

---

[7] In comparing the growth performance of the EU with the US during the 1990s it is important to take into account the fact that the EU and the US experienced a different timing in their business cycles during the 1990s. We find that the US experienced a recession during 1990–1, whereas the recession in the EU occurred during 1992–3 (see Fig. 5.19 in the previous chapter). We can conclude that the different timing of the recessions in the 1990s should not affect the difference in the average growth rates during that period observed between the US and the EU.

Table 6.2. Average growth rates of GDP in the
EU and the US (1980s and 1990s)

	EU	US
1981–90	2.4%	2.5%
1991–6	1.5%	2.3%

Source: EC Commission, *European Economy.*

debts and deficits, while at the same time they should reduce their inflation rates. Applied by many countries at the same time, this lead to strong deflationary forces. First, when all countries reduced their budget deficits by reducing spending and/or increasing taxes the negative effects on aggregate demand in one country spilled over to the other countries.[8] As a result, economic activity in these countries reduced, thereby increasing the budget deficits in the same countries and forcing the authorities to intensify their attempts to reduce budget deficits. Thus, the simultaneous application of the same restrictive budgetary policies may have contributed to the low-growth environment in the European Union, which in turn made budget cutting exercises less effective.

Second, the inflation convergence requirement forced monetary policies in all countries participating in the Maastricht strategy to be restrictive. This is illustrated in Fig. 5.19 in the previous chapter. We observe that the EU countries maintained their short-term real interest rates at a high level during the recession of 1992–3 and generally kept these real interest rates higher than in the US. The latter country allowed its real interest rate to drop significantly during the recession (which occurred in 1990–1). Together with the restrictive budgetary policies, the restrictive European monetary policies reinforced the deflationary dynamics in the Maastricht countries. In this process of monetary restriction, another spillover effect between countries has been at work. Given their commitment to fixed exchange rates, the Maastricht countries only have one instrument to improve their competitive position relative to the other countries, i.e. by reducing their inflation rates below the level of these other countries. As a result, a process of competitive disinflation was set in motion in which countries tried to outdo the others in reducing inflation.[9]

[8] Recently, some evidence has been provided by Giavazzi and Pagano (1990) that policies of budgetary restriction do not have to lead to a reduction of aggregate demand. In particular, in countries with a high government debt, forceful policies of cutting budget deficits may create favourable expectational effects that lead to a significant reduction of the domestic interest rate. This may then compensate for the unfavourable Keynesian demand effect of restrictive fiscal policies. These exceptional effects will probably be small for most EU countries, for two reasons. First, the expectational effect works best in countries with a high initial government debt (e.g. Belgium, Italy). In countries with relatively low initial debt (France, Germany) this effect is likely to be small. Second, the expectational effect works well if the budget cutting exercise is done all at once, and not in a gradual way. Political factors explain why most governments have chosen the gradual approach, thereby very much limiting the favourable effects of expectations.

[9] This policy of competitive disinflation was followed quite actively by France. Trichet, the governor of the Banque de France, openly admitted that this was France's objective. See Connolly (1995).

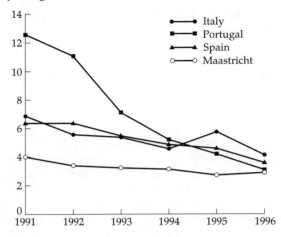

Fig. 6.2. Inflation rates and the Maastricht norm

Those countries who did not follow this policy of competitive disinflation were nevertheless forced to do so because the Maastricht inflation norm became a downward-spiralling target for these countries. In Fig. 6.2 we show some evidence substantiating the previous analysis. We selected the Southern European countries most eager to join the monetary union. We observe that they have made dramatic efforts to reduce their inflation rates. As a result, these countries would have reached the inflation norm in 1996 had this norm remained at the level achieved at the signing of the Treaty in 1991. At the same time, the Maastricht norm moved downwards from more than 4% in 1991 to less than 3% in 1996.

Thus, the policy mix of monetary *and* budgetary restriction applied by many countries may have had a self-defeating aspect. Not only did this policy mix reduce the effectiveness of measures to reduce government budget deficits, it probably explains why the government debt actually increased as a per cent of GDP. We show the evidence in Fig. 6.3. We observe that, after 1993, EU countries managed to reduce their budget deficits on average. However, at the same time the debt-to-GDP ratio continued to increase, reaching 70% in 1996. This striking development is undoubtedly related to the policy mix of budgetary and monetary restriction, which has led to a situation in which the budget cutting policies have been performed in an environment of low nominal growth of GDP. Remember that the debt ratio is the ratio of government debt to GDP. Thus when budget deficits decline, the *growth rate* of the numerator (the debt) declines. At the same time the deflationary policy mix reduced the growth rate of the denominator (the GDP). The latter effect compensated for the former, preventing the debt ratio from declining.

The difficulties of achieving the budgetary targets the Maastricht countries set themselves contrast with the success of the US, which managed to reduce its budget deficit and debt level during the same period. This success was certainly

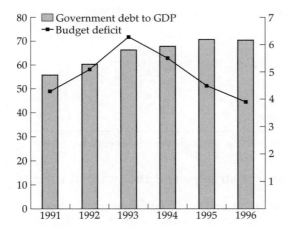

Fig. 6.3. Average government debt-to-GDP ratio and budget deficit in the EU

Source: European Commission, *European Economy*, 1996.

helped by the fact that the US policy mix was quite different. The budgetary restriction that was applied was combined with monetary ease. This contributed to stronger economic growth than in the EU, which encouraged the implementation of restrictive budgetary policies. At the same time the policy mix of monetary ease and fiscal restriction helped to keep inflation in check. In 1996 the rate of inflation in the US was below the average inflation rate in the EU.

One can conclude that the Maastricht-imposed policy mix of fiscal and monetary restriction not only contributed to a lot of economic hardship in Europe: more importantly it had a self-defeating aspect, which explains why many countries were further from the targets in 1996 than in 1991 when the treaty was signed. Whether a sufficient number of countries will achieve these targets in time is difficult to predict, as it depends on the degree of flexibility with which these will be interpreted. Ultimately, the willingness to take a flexible stance will depend on the degree of political commitment countries have towards monetary union. If this commitment is strong enough, the required flexibility will be forthcoming and EMU will start on time.[10] If not, the start of EMU in 1999 will be made very difficult.

---

[10] It should be stressed that the Maastricht Treaty allows for a flexible interpretation of the convergence criteria. The budgetary criteria are not rigid. The Treaty allows for temporary deviations from the budgetary norms and also allows for deviations due to exceptional circumstances. In addition, based on the reports of the Commission and the EMI, the Council will decide for each country 'whether it fulfils the necessary *conditions* for the adoption of a single currency' (art. 109j–2; italics mine). It is significant that the Treaty uses the word *conditions* instead of *criteria* (which will be analysed in the Commission and EMI reports). This suggests that the Council can make a different evaluation than the Commission and the EMI. The latter have to base their evaluation of a country's readiness to join EMU on its progress towards meeting the convergence *criteria*.

Table 6.3. Government budget deficits in the EU and the US (per cent of GDP)

	EU	US
1991	4.3	3.5
1992	5.1	4.5
1993	6.3	3.4
1994	5.5	2.0
1995	4.5	1.6

Source: European Commission, *European Economy*, 1996.

3.2. The Maastricht strategy and the unity of the EU

The Maastricht inflation and budgetary convergence requirements carry the risk of splitting up the European Union. In this section we elaborate on this theme. We first concentrate on the inflation and then on the budgetary convergence requirement. The problem we will analyse in this section arises mainly because countries with a poor reputation concerning inflation and budgetary policies may get trapped into a dynamics making it difficult to converge precisely because their initial reputation is poor.

The dynamics of inflation convergence

In order to analyse the problem of inflation convergence we will use our two-country Barro–Gordon model discussed earlier. Let us assume that Germany and Italy agree that the Italian inflation rate should converge towards the low German inflation rate before the step towards full monetary union is taken. As a result, the Italian authorities announce a policy of gradual reduction of the inflation rate. They also announce that they will keep the exchange rate with the German mark fixed.

We represent the Barro–Gordon model for Italy in Fig. 6.4.[11] Prior to the announcement of the policy of gradual convergence, equilibrium was located at point E. The announcement of the new policy can be interpreted as a change in priorities of the Italian authorities. They are now willing to accept more (temporary) unemployment in order to reduce inflation. We represent this change in policy preferences by the indifference curves I'_1, I'_2, which are flatter than the initial ones. The problem, however, is that economic agents are uncertain about the seriousness of this 'conversion'. As a result, they will not be willing to believe this announcement. They will want proof. This can only be obtained from observed facts. Economic agents will be willing to believe the seriousness of the authorities only after they have observed that the authorities have implemented disinflation policies and that they are willing to let unemployment increase. As a result, it will be necessary for the economy to move to

---

[11] In fact we use the Barro–Gordon model here, as it was extended by Backus and Driffill (1985).

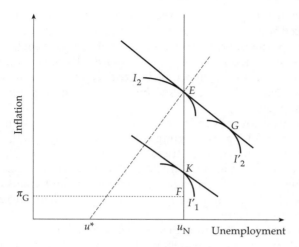

Fig. 6.4. The cost of gradual disinflation in Italy

point G. At that point, the Italian authorities reveal their changed preferences, i.e. the slope of the short-term Phillips curve is equal to the new indifference curve. Thus economic agents observe from the facts that these preferences have changed and that the authorities are now willing to let unemployment increase more in order to achieve a given reduction of inflation.

This situation will gradually lead to a change in expectations, allowing the short-term Phillips curves to shift downwards. We move to point K. It follows that this transition process will be a slow and costly one. It should be noted that this is the strategy that countries like Italy, France, Portugal, and Spain have applied from the second half of the eighties on. As was documented in the previous chapter, this disinflation process was slow and painful.

Some economists have argued that this strategy of slow convergence of the inflation rate in the high-inflation countries to the German level is risky, and in the end may fail to bring the inflation rates to equality. The reason for drawing this pessimistic conclusion can be explained as follows. It is unlikely that the new Italian policies will convince economic agents that the Italian authorities are as 'hard-nosed' about inflation as the German authorities. Therefore, the new inflation equilibrium (in K) will still be higher than in Germany (see Fig. 6.4). Thus, the Italian inflation rate will not completely converge to the German one, but will hover above it.

This situation leads immediately to a credibility problem. As the Italian rate remains above the German one, the price levels of the two countries will tend to diverge continuously. This leads to a continuous loss of competitiveness of Italian industry (a real appreciation of the Italian lira). In the end this situation becomes unsustainable. The option of a devaluation will have to be exercised. Convergence will be all the more difficult thereafter.

In the previous chapter we showed evidence that the Italian and Spanish economies faced this problem during 1987–92. This made the devaluation of the lira and the peseta in September 1992 inevitable. It also suggests that a slow transition process, which makes the steps towards the final stage of monetary union conditional on the existence of full convergence of inflation rates, is a dangerous one. Not only is it a slow one; if stretched out too long it may become unsustainable.

Recently progress towards inflation convergence of the high-inflation countries, although painful, has been quite substantial (see Fig. 6.2), so the possibility cannot be excluded that these countries will meet the Maastricht inflation criteria before the start of EMU. A failure to be admitted (because of unfavourable budgetary performance), however, would almost certainly set these countries back, as the ensuing devaluations of their currencies would ignite inflationary pressures.

Let us now turn our attention to the question of budgetary dynamics. We first focus on the interaction between inflation and budgetary convergence and then between interest rate and budgetary convergence.

The dynamics of inflation and budgetary convergence

Countries that start a process of budgetary restriction aimed at reducing the government debt will find that this can be made difficult if at the same time they try to reduce their inflation rate and lack the credibility to do so. This can be explained as follows. Take a country with a high debt-to-GDP ratio. It also has a high inflation rate to start with, and starts a policy of disinflation. If this policy lacks full credibility the *expected* inflation rate will not (or only slowly) decline. Typically, the observed inflation rate declines faster than the expected one. This was the case of Italy, discussed in the previous section. This creates the following problem for the public finances of that country. As expected inflation does not decline the nominal interest rate (which reflects prevailing inflation expectations) does not decline either. The observed inflation rate, however, declines so that the ex-post real interest rate (i.e. the nominal interest rate − the observed inflation rate) increases. The latter then increases the burden of the debt. This will force the government either to raise taxes or reduce spending so as to prevent the debt-to-GDP ratio from increasing. We conclude that debt reduction policies are made more difficult when governments engage in disinflationary policies that lack credibility. The interested reader may look at Box 9, where the preceding analysis is presented more formally.

Box 9. Debt reduction and disinflation

This argument can be presented more formally in the following way. Let us start from the government budget constraint:

$$G - T + rB = dB/dt + dM/dt \qquad (B9.1)$$

where G is the level of government spending (excluding interest payments on the government debt), T is the tax revenue, r is the interest rate on the government debt, B, and M is the level of high-powered money (monetary base).

The left-hand side of equation (B9.1) is the government budget deficit. It consists of the primary budget deficit ($G - T$) and the interest payment on the government debt (rB). The right-hand side is the financing side. The budget deficit can be financed by issuing debt (dB/dt) or by issuing high-powered money (dM/dt). We will assume, however, that monetary financing constitutes such a small part of the financing of the government budget deficit in European countries that it can safely be disregarded. (In this connection see Table 1.2 which shows how small monetary financing has become in the EU countries.) In the following we represent the changes per unit of time by putting a dot above a variable, thus $dB/dt = \dot{B}$.

It is convenient to express variables as ratios to GDP. Let us therefore define

$$b = B/Y \qquad (B9.2)$$

where Y is GDP, so that b is the debt-to-GDP ratio.

This allows us to write

$$\dot{b} = \dot{B}/Y - B\dot{Y}/Y^2 \qquad (B9.3)$$

or using (B9.2) and rearranging

$$\dot{B} = \dot{b}Y + b\dot{Y} \qquad (B9.4)$$

Substituting (B9.4) into (B9.1) yields

$$\dot{b} = (g - t) + (r - x)b + \dot{m} \qquad (B9.5)$$

where $g = G/Y$, $t = T/Y$, $x = \dot{Y}/Y$ (the growth rate of GDP), and $\dot{m} = \dot{M}/Y$. Given our previous assumption, $\dot{m} = 0$.

We can write x as follows:

$$x = \pi + q \qquad (B9.6)$$

where π is the rate of inflation and q is the real growth rate of GDP. In order to focus on the role of inflation in the process of debt reduction we will set $q = 0$. This we do only for the sake of simplicity: it does not affect the analysis. Equation (B9.5) can therefore be rewritten as

$$\dot{b} = g - t + (r - \pi)b \qquad (B9.7)$$

We will now invoke the Fisher relation, which defines the nominal interest rate as follows:

$$r = \rho + \pi^e \qquad (B9.8)$$

where π^e is the expected inflation rate, and ρ is the real interest rate.
Substituting (B9.8) into (B9.7) yields

$$\dot{b} = g - t + (\rho + \pi^e - \pi)b \qquad (B9.9)$$

We observe that it is only the unanticipated component of inflation ($\pi^e - \pi$) that affects the budget constraint, i.e. inflation which is higher than expected lowers the debt burden. Conversely, inflation that is lower than expected raises the debt burden (increases the debt-to-GDP ratio). Fully anticipated inflation ($\pi = \pi^e$) does not change the debt burden. The reason is that a fully anticipated inflation reduces the real value of the outstanding debt and increases the nominal interest rate at the same time.

We can now use (B9.9) to analyse the problem of a country which starts the Maastricht convergence strategy with an unfavourable inflation reputation. As argued earlier this country will find it difficult to reduce inflationary expectations, as agents do not fully trust the announced strategy. This is the reason why in the Barro–Gordon model we presented the high-inflation country (Italy) is forced to move downwards along the short-term Phillips curve. During this process observed inflation declines below expected inflation. This is quite important because as can be seen from equation (B9.9), this will lead to an increase in the debt-to-GDP ratio, b. Put differently, the lack of credibility of the disinflationary process raises the (ex-post) real interest rate ($\rho + \pi^e - \pi$) thereby increasing the debt burden of the government. The authorities must therefore increase taxes or reduce spending just to prevent the debt-to-GDP ratio from increasing. Thus, the inflation convergence requirement makes debt reduction more difficult when, as in the case of Italy, a credible anti-inflation policy is difficult to follow.

Interest rate convergence and debt dynamics

The previous problem is compounded by the interaction between the interest rate convergence requirement and debt dynamics. This can be described as follows. When speculators have doubts that the high-inflation country (Italy) will be accepted into the EMU, they will incorporate a devaluation premium into the long-term bond rate. These doubts may arise from the doubts concerning inflation convergence. In that case we have the same problem as the one we have just analysed. It is, however, also possible that these doubts arise from a belief that the Italian authorities will not manage to reduce the budget deficit. In that case the devaluation premium, which is built into the long-term bond rate, will make budgetary convergence more painful and more difficult.

As was mentioned earlier the interest rate convergence criterion has a self-fulfilling aspect. The self-fulfilling nature of the interest rate convergence requirement is reinforced by the interaction with budgetary convergence. If the market believes that a country will not be accepted into EMU, this raises the long-term interest rate. This in turn increases the government budget deficit and the debt-to-GDP ratio, taking the country further away from the budgetary norms, thereby validating the expectation that the country will not

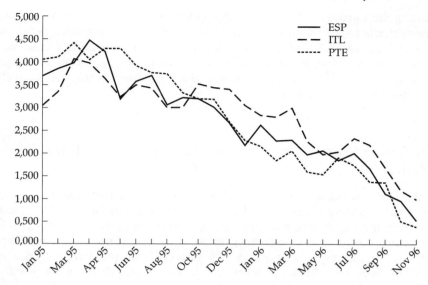

Fig. 6.5. Spread of long-term bond rates between ESP, PTE, ITL and DM

Source: J. P. Morgan.

be admitted. In this process, the country may be pushed into a 'bad equilibrium' of increasing debt and deficits.

Given the self-fulfilling nature of the interest rate criterion, however, the reverse is also possible, so that a 'good equilibrium' becomes possible. This may in fact have happened during 1995–6. As a result of the increasing political commitments of the governments of Italy, Portugal, and Spain to join EMU and their willingness to introduce tough budgetary measures, the markets have become optimistic about the entry of these countries into the EMU on 1 January 1999. This has led to a dramatic drop in the long-term interest rates in these countries. Fig. 6.5 shows the size of this drop. We present the spread between these countries' long-term bond rates and the German bond rate. At the end of 1996 this spread had declined to less than 100 basis points. This is bound to make budgetary convergence easier, as the reduction in interest payments reduces the government budget deficits.

The risk implicit in this self-fulfilling dynamics of convergence is that a change in the market's evaluation of the probability of entry by an individual country can move this country back into a bad equilibrium of high interest rates, high interest payments leading to high deficits and a renewed budgetary divergence, making it impossible for the country to join EMU. It cannot be excluded that such a sudden reversal occurs for a number of countries with weak public finances as we approach the date of entry into EMU.

We conclude that the Maastricht convergence requirements carry the risk of splitting the European Union apart. If applied strictly, the Maastricht convergence criteria may ban some countries from entry for a prolonged period of time and against their wishes. This is bound to create political and economic conflicts between the 'ins' and the 'outs'. Paradoxically, therefore, the Maastricht treaty, which aimed at deepening the European Union, may actually contribute towards disunity.

3.3. Convergence requirements as guarantees for low future inflation

In the previous sections we argued that the Maastricht convergence strategy has a self-defeating character and risks splitting the European Union apart. One may interject here that all this may be true, but that there is no other way to organize a transition to monetary union in Europe. If this future monetary union is to be one that produces low inflation, the prior convergence of inflation, debt, and deficits is necessary (as was argued earlier when we discussed the rationale of the Maastricht convergence requirements). If that is so, we will have to live with the fact that the transition must necessarily be painful and risky.

There is, however, a major problem with this interpretation. Countries going through the convergence game may act opportunistically, i.e. they may do this today so as to gain access later.[12] Once they are in the union, they will reveal their true preferences. In addition, even if the present governments of the candidate member countries are serious about inflation and submit themselves whole-heartedly to the disinflationary process, they do not commit future national representatives in the ECB to the same monetary policy stance. To give an example: the fact that France has reduced its inflation rate during the 1990s so as to be accepted into the monetary union will not bind the French representative in the ECB in the year 2010 to follow the same low-inflation policies. At that moment the policies of the French authorities of the 1990s will have become irrelevant. The only relevant constraint in 2010 will arise from the nature of the monetary institutions that have been created and the incentives these institutions will give to the European monetary authorities to produce low inflation. Thus, one can conclude that the entry requirements formulated in the Maastricht Treaty in no way solve the German problem. They provide no additional guarantee that the future monetary policies of the ECB will be geared towards low inflation (additional to the guarantee enshrined in the statutes of the ECB).

We can also formulate the previous analysis in the context of game theory. The convergence dynamics can be seen as a game in which countries are

---

[12] Recent examples of such opportunistic behaviour are provided by budgetary policies. Increasingly, as we move towards the EMU entry date, countries start window-dressing their government budgets.

rewarded by following painful policies of monetary and budgetary restriction. Failure to reduce inflation and budget deficits carries a harsh punishment, i.e. exclusion from EMU. This gives countries strong incentives to comply and to institute disinflationary strategies (at least those countries who deem the permanent benefits of joining EMU to exceed the temporary costs of a disinflationary strategy). Once these countries are admitted into the EMU, however, the nature of the game, i.e. its reward–punishment structure, changes fundamentally. Suddenly the punishment vanishes.[13] This is likely to lead to a relaxation of the budgetary policies of these countries.

The realization that this is a serious problem of the Maastricht convergence criteria has led the German Minister of Finance, Theo Waigel, to propose a 'stability pact' that would be implemented after the start of EMU. It was adopted at the December 1996 Summit Meeting in Dublin. The idea is very simple: since after the start of EMU the punishment for bad budgetary behaviour disappears, a punishment structure that will apply after 1999 should be instituted. Countries that fail to meet the 3% budgetary norm during a given period of time would be punished by having to pay a fine ranging from 0.2 to 0.5% of GDP.[14] The original German proposal was that these fines would be applied automatically if the deviations from 3% are maintained for a year or longer. The stability pact agreed upon at Dublin has dropped the automatic application of these fines. These fines will be applied after deliberation in the Council, unless the excessive deficit is due to exceptional circumstances (e.g. a deep recession involving a drop of GDP of more than 2%). We will return to the problems that the application of this stability pact could lead to in a future EMU when we discuss fiscal policies in a monetary union (Chapter 9).

4. Alternative transition scenarios

As mentioned earlier, the historic evidence suggests that the Maastricht transition strategy is not the only one available. Successful monetary unions have been created after a quick transition and without prior convergence requirements. In this section, we show that such an alternative approach is possible

---

[13] This is not strictly true. The Maastricht Treaty formulates a set of penalties that countries will be subjected to if they are found to have 'excessive deficits' (i.e. too large a budget deficit relative to the 3% norm and a government debt that does not decline sufficiently fast towards the 60% norm). The first penalty would be a condemnation, the second a ban on obtaining credits from the European Investment Bank. The Maastricht Treaty also mentions the possibility that countries may have to pay fines if they persist in having excessive deficits. The Treaty, however, is not specific on this. As a result, there is a general feeling that these penalties are not very strong and may not deter individual countries from accumulating excessive deficits.

[14] This was not really new since the Treaty has a provision making it possible for fines to be levied on countries with excessive deficits. The new feature of the stability pact was that this possibility was given a practical content.

and that paradoxically it can facilitate the convergence of inflation and budgetary policies.

Let us return to the inflation convergence requirements first. We use the Barro–Gordon model again. Let us now assume that instead of requiring Italy to reduce its inflation rate prior to entry, we announce that both countries have agreed to form a monetary union and that the Italian authorities accept the idea that the future European Central Bank will be run like the Bundesbank, so as to ensure that the future EMU will produce low inflation (what exactly this means in terms of the institutional constraints to be imposed on the future ECB will be discussed in Chapter 8). In addition, we announce that the monetary union will be organized quickly. In other words, we announce a monetary reform to be organized, say, in six months. At the moment of the monetary reform, say on E-day, the lira disappears and is replaced by the euro, which is managed by tough central bankers in Frankfurt.

It can now easily be seen that there is a major advantage for Italy if it proceeds in this way. We show this in Fig. 6.4. Prior to E-day, the inflation equilibrium in Italy was given by point *E*. After E-day, Italy can immediately move to point *F*, the European (German) inflation rate, without any cost in terms of unemployment. The euro (a money issued by the ECB) is now circulating in Italy, and will be subject to the European inflation rate. Economic agents realize this, and are willing to quickly reduce their expectation about the inflation rate that will prevail in Italy. Thus, Italy can profit from the better reputation of the European Central Bank.[15] By taking over the money and the institutions of the low-inflation country, Italy can capture a large welfare gain from the monetary union. We conclude that there is a great advantage to shock therapy (a monetary reform). Note also that in this shock therapy approach there is no need for Italy to reduce its inflation rate *prior* to the formation of the monetary union. The monetary reform does the trick of bringing the Italian and the German inflation rates together.

Contrasting this shock therapy (without prior inflation convergence) with the Maastricht inflation convergence strategy, we can say that the former is likely to be much less painful than the latter. Paradoxically, the quick approach succeeds in forcing the convergence of inflation more quickly than the Maastricht convergence strategy. This is not really surprising. Inflation is the loss of purchasing power of money. Thus, when inflation is higher in Italy than in the rest of Europe, we really say that the lira is losing its purchasing power faster than the other moneys. A monetary reform is a change of regime which eliminates the lira, and thus the loss of purchasing power associated with that currency. The euro takes over in Italy, so that the Italian inflation rate will essentially be the loss of purchasing power of the euro, managed from

---

[15] It can be argued that at least initially the ECB will not have the same low-inflation reputation as the Bundesbank. As a result, the gain in Italy's reputation wil be somewhat reduced. Nevertheless, the reputation of the ECB is likely to be greater than that of the Italian monetary authorities, so that the monetary union will still make it possible for Italy to acquire a better inflation reputation than if it stays out of the union.

Frankfurt, much as the rate of inflation in Bavaria is pretty much the same as in Schleswig-Holstein because in the two regions the same Deutschmark is used.

It should be pointed out that complications may arise if there exists inertia in the prices. This could be the case if, for example, wages are indexed to the price index, or if wages continue to increase according to some formula agreed upon prior to the monetary reform. In order to make the shock therapy successful these contracts will have to be changed at the time of the monetary reform. Failure to do so could lead to divergent price and wage movements in the countries participating in the union, at least during the early phase of the monetary union.

Two points should be stressed concerning the previous argument. First, when inflation rates are very different prior to the start of the monetary union, the transitional problems due to price inertia become more pronounced. Thus, some convergence of inflation prior to EMU is desirable. The criticism levelled here against the Maastricht inflation convergence is that it is excessively restrictive. Put differently, we argue that EMU could have started among the EU countries in, say, 1996 with the then existing differences in inflation (which amounted to at most 3%) without creating significant problems. Second, the criticism formulated here is quite different from the arguments developed in the chapters about the costs of a monetary union. There we argued that structural and institutional features can lead to different wage–price spirals in different countries. If these countries then join in a monetary union they may be stuck in unsustainable situations. These countries should therefore not form a monetary union or if they do, they should change these institutions. Prior convergence of inflation as required in the Maastricht Treaty does not deal with this problem.

In the previous sections we argued that the easiest way to force inflation convergence consists in starting a monetary union. In a similar way it can be argued that the start of EMU without prior budgetary convergence would actually facilitate budgetary convergence. As we saw earlier, the fact that a country is expected to be left out of EMU increases the debt burden and intensifies the efforts the government of that country will have to exert to reach the Maastricht budgetary norms. Allowing the country into the union can facilitate the reduction of the government budget deficit because it is likely to lead to a significantly lower interest burden. Thus, allowing countries with weak public finances into the EMU facilitates their convergence. Whether this implies imposing post-entry budgetary rules (like the 'stability pact') is another issue, which will be discussed in Chapter 9. The point is that the very fact that countries with a high debt are allowed into EMU creates a significant budgetary bonus, facilitating achievement of the 3% budgetary norm.

5. The political economy of monetary union in Europe

Our criticism of the Maastricht transition strategy seems devastating. Not only has it brought hardship to Europe, it may actually lead to a situation where many EU countries that are eager to join may not reach the Maastricht norms in time. In addition, it risks splitting the European Union apart. What is more, even if countries meet the criteria and join EMU, it does not even give good guarantees to the low-inflation country (Germany) that the fulfilment of the criteria will bring about low inflation in the future EMU. Finally, alternative and less costly transition strategies are available. Given such handicaps, one wonders why the drafters of the Maastricht Treaty chose this particular strategy. In order to answer this question it is useful to discuss the political economy of EMU.

The starting-point of such a political economy analysis could again be the Barro–Gordon model as discussed in Fig. 6.1. We concluded from that analysis that Germany has nothing to gain (in terms of inflation and unemployment) from joining a union with high-inflation countries. Even if it insists, and succeeds, in obtaining a European central bank which is a copy of the Bundesbank, Germany still does not gain anything. It will face the same inflation equilibrium. All the gains are for the high-inflation countries. (Note, however, that countries may have other reasons to join a monetary union. For example, they may consider the other (efficiency) gains of a monetary union, which we discussed in Chapter 3, to be significant enough to outweigh the welfare loss resulting from an increased risk of inflation. Or they may have political objectives for pursuing monetary union.)

The preceding discussion is important for understanding the political economy of monetary integration in Europe. This is influenced by the poor incentives for Germany to be part of a full monetary union. In fact, the incentive structure of Germany is such that it is in its interest to keep the union small, and to prevent high-inflation countries from belonging to the union. We illustrate this using the Barro–Gordon model again. We now consider three countries, Germany, France, and Italy. In order to simplify the analysis we assume that these three countries are identical except for the preferences their authorities attach to combating inflation and unemployment. We represent these three countries in Fig. 6.6. The short-term Phillips curves and the natural unemployment levels are the same for the three countries. The authorities' indifference curves concerning the choice between inflation and unemployment differ. Germany has flat indifference curves reflecting the high weight it attaches to reducing inflation. Italy has steep indifference curves because it attaches less weight to low inflation. France has an intermediate position. The equilibrium inflation rates in the three countries are obtained in points *A*, *B*, and *C*, with Germany having the lowest and Italy the highest inflation rates.

A monetary union implies that a common rate of inflation will have to be

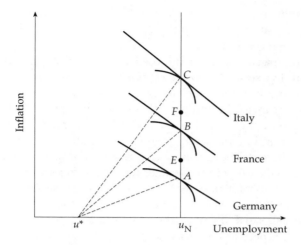

Fig. 6.6. Inflation equilibrium with three countries

selected. Assume now that in the future monetary union the European Central Bank will reflect the preferences of those countries that are members of the union. Thus, the inflation outcome in the union will be determined by the size of the union. If Germany and France form a union without Italy, the inflation equilibrium will be located between points *A* and *B*, say in *E*. This union will be attractive to Germany if it considers the efficiency gains of the union to outweigh the loss in welfare due to the higher rate of inflation.

If, however, the union consists of the three countries, the inflation equilibrium of that large union will be located between points *A* and *C*, say at *F*. Thus, the loss of welfare for Germany due to the higher rate of inflation will be larger than in the smaller union. In order to induce Germany to be part of the larger union, the perceived efficiency gains of the union must be correspondingly larger. From the point of view of Germany the larger union is less likely to be attractive than the smaller union with France only.

This leads to the conclusion that Germany's interest may very well be to keep the European monetary union relatively small. This conclusion is reinforced by the fact that during the 1980s and the 1990s Germany achieved a dominating (some will say hegemonic) position in monetary policy-making in Europe. A monetary union necessarily means that the power of the German monetary authorities will be diluted. The German central bankers will not be the only ones sitting around the table in Frankfurt. Thus, a monetary union means a loss of power for Germany and a corresponding gain in power for the other EMU members to decide monetary affairs in Europe. It is, therefore, understandable that the German monetary authorities are looking for ways to limit their loss of power in the future EMU. One way to achieve this is by keeping the union small. It also explains the insistence of the German authorities that only countries satisfying the convergence criteria will be able to be part of the monetary union.

Thus, the convergence criteria imposed by the Maastricht Treaty are not conditions that for some scientific economic reasons must be satisfied if countries want to join a monetary union. As was argued earlier, a monetary union could be organized in Europe with a shock therapy, i.e. without imposing the requirement that inflation rates and interest rates have to converge before countries join in a monetary union. The function of these convergence criteria is very different. They are necessary to keep the size of the union small. In so doing, they take care of the German reluctance to join a monetary union with too many members. For as we have seen, these convergence criteria have been formulated in such a way as to make it unlikely that all the EU members wishing to join will satisfy them in the near future.[16]

The surprising aspect of the analysis of the political economy of monetary union is that it leads to a conclusion similar to that of the analysis of optimum currency areas (OCAs). In the latter case we arrived at the conclusion that a monetary union with fifteen EU members is probably suboptimal from a welfare point of view. The political economy analysis of the Maastricht Treaty leads to the conclusion that a monetary union with fifteen members is unlikely to occur, because it is suboptimal for the leading country in the EU and therefore will be resisted by that country.

Both the OCA theory and the political economy analysis of the Maastricht Treaty also come out in favour of a small monetary union in Europe. The criteria to select the small group of countries, however, are different. In the OCA theory criteria such as wage flexibility, labour mobility, and degree of economic integration stand out as the important ones for determining whether a country should be part of the union.

In the Maastricht strategy these criteria of the OCA theory play no role at all. Instead, nominal convergence criteria (together with budgetary convergence) are to decide which countries will join the union. These criteria are there not because of economic necessity, but mainly because of the need to satisfy the interests of Germany.

While Germany will use the convergence criteria to pursue its national interest, other countries will also do this. And the overriding interest of the high-inflation countries like Italy and Spain is to get early access to the monetary union. This can be seen from the same Fig. 6.6 and from our previous discussion. These countries stand to gain a lot from being accepted into the union from the start. For these countries the convergence criteria will be used as stepping stones to achieve access. A major conflict of interest will erupt between Northern European countries wishing to keep the union rela-

---

[16] At least if they play the Maastricht game correctly. The pressure to 'fudge' the budget numbers, however, will be strong. We will come back to this problem.

tively small and Southern European countries wishing to be included. This conflict will have to be resolved.

The way this conflict will be resolved depends on two factors. First there is the decision-making process. In this connection, it is important to realize that the decision to allow individual members into EMU will have to be made by qualified majority (Maastricht Treaty, Article 109*j*-4). Second, there is the increasing likelihood that many countries, including several of the hard-core Northern group, will not meet all the Maastricht conditions. The decision-making process on the membership issue, therefore, could lead to a deadlock. The reason is the following. The qualified majority rule implies that a limited number of countries (say three or four) can use a blocking minority.

In order to illustrate this Table 6.4 gives the distribution of the votes in the Council of Ministers. It can immediately be seen that there is a wide scope for coalition-building by the losers in the Maastricht game. In other words, the countries that are commonly considered to form the hard-core and want to start the currency mini-union in 1999 (Germany, France, Benelux, Austria, Ireland) will need the approval of the 'losers', who can form various coalitions to block the entry of one or more of the hard-core countries. The question then is whether there will be objective grounds for the losers to exert their negative voting power. The answer is most likely positive.

The probability that one or more of the hard-core countries will not satisfy one or more of the Maastricht criteria now looks quite likely. The deficit and/ or debt criteria, for example, are almost certainly not going to be met by several hard-core countries. Even Germany may get into trouble with these criteria. In order to let these hard-core countries pass the Maastricht entrance exam, therefore, it will be necessary to give a sufficiently flexible interpretation to the Maastricht criteria, or alternatively, the numbers will have to be manipulated. (The latter seems to be the approach France has taken in its 1997 budget proposal. The French government plans to write into its budget a large cash payment to be made by France Telecom as current revenue, in exchange for the government taking over the future pension liabilities of the personnel of France Telecom. The present value of these future liabilities is kept outside the budget. Thus, the future French government is reducing today's deficit at the expense of larger budget deficits in the future.) This, however, will open the door to similar interpretations and/or manipulations of other Maastricht criteria, which may unravel the whole Maastricht convergence game.

The other countries, most of them Southern European, will have good grounds to oppose the selective use of the Maastricht criteria to allow the 'core' countries into the EMU. The Maastricht Treaty gives them the power to block such a scenario. These countries will have a strong incentive to trade their votes in exchange for a similar benevolent judgement concerning their convergence efforts.

The conflict that we have described then leads to the following view about the choices that the EU countries will face. One is to start EMU with a

Table 6.4. Distribution of votes in the Council of Ministers

Germany	10
France	10
Italy	10
UK	10
Spain	8
Belgium	5
Greece	5
Netherlands	5
Portugal	5
Austria	4
Sweden	4
Denmark	3
Finland	3
Ireland	3
Luxemburg	2
Total	87
Qualified majority	62
Blocking minority	26

relatively large group of countries (including the Southern European countries). This will make the latter happy but will make Germany unhappy. In fact Germany might be so unhappy with this idea that it would prefer a second option, which is the postponement of the start of the monetary union. What about the third option, i.e. starting EMU with a core group of countries (mainly Northern European countries)? It is difficult to see how Southern European countries, which by the qualified majority rule will have the power to block this scenario, will accept this. The only way this scenario, therefore, can become reality is by giving sufficiently strong guarantees to the countries left out that they will be allowed into the union quickly. We will discuss the nature of these guarantees in the next chapter, where we analyse the relationships between the 'ins' and the 'outs'.

A final problem should be mentioned here. As was argued earlier, the link between budgetary restrictions and monetary union implicit in the Maastricht Treaty certainly has made it easier for governments to impose unpopular budgetary measures. At the same time this link has introduced increasing scepticism in the population about the desirability of EMU. Many citizens who have been told by their respective governments that monetary union would increase the welfare have instead been confronted by increasing taxes and less government spending in the name of the same monetary union. It is no surprise that they increasingly distrust the idea that monetary union is a good project, and that the opposition to EMU is increasing. In the end the opposition may gather enough momentum to weaken the political resolve to go ahead precisely at the moment resolve will be most needed.

6. Conclusion: Suggestions for a different approach

How can one avoid the stalemate that is likely to arise in 1998? In order to answer this question we have to solve the paradox that is inherent in the Maastricht strategy. On the one hand the entry into the union of countries with a weak reputation concerning inflation and/or public finances is perceived by Germany to go against its national interest because it could jeopardize price stability in the future union. On the other hand letting these countries into the union will make it easier for them to reduce their inflation and their government debt. At the same time allowing these countries into the union would eliminate the risk of a deep division of the European Union. How can this paradox be solved? The analysis of this chapter suggests a number of reforms in the transition process towards monetary union. The general principle that should guide this reform can be formulated as follows. The transition to EMU should put less emphasis on entry requirements and more on strengthening the future European monetary union so that the authorities can deliver on their mandate to produce price stability. There are several ways in which this can be achieved. First, the political independence of the future European Central Bank must be ensured. This principle is in fact already enshrined in the Treaty of Maastricht and will be discussed in greater detail in Chapter 8. Second, additional institution-strengthening devices will have to be designed. In this connection it has been proposed to allow only the representatives of the countries that satisfy the budgetary norms to vote in the European Central Bank. This and other proposals on the design of the ECB will be discussed in Chapter 8. Third, rules on the conduct of budgetary policies in the post-EMU period must be implemented. The stability pact that was approved at the Dublin Summit Meeting of December 1996 does exactly that. We will discuss this stability pact in Chapter 9 when we study fiscal policies in monetary union.

It is important to see these institution-strengthening measures and rules as a quid pro quo. By strengthening the institutions of the future EMU the German public can be convinced that the future monetary union will provide for low inflation. This then makes it possible to relax the entry requirements (which, we have argued, provide few guarantees to Germany). The relaxation of the entry requirements then reduces the risk that the European Union will split into two parts, producing great economic and political strains.

7

TECHNICAL PROBLEMS DURING THE TRANSITION PERIOD

Introduction

The problems discussed in the previous chapter are quite formidable and may jeopardize the start of EMU. Let us, however, assume that these problems will be overcome and that EMU begins on time with a significant number of countries. Given the commitment of major EU countries towards EMU, this is not an unreasonable assumption to make. A number of technical problems will then arise and will have to be resolved.

1. How to fix the conversion rates?

The first problem arises because of the uncertainty that will exist concerning the choice of the conversion rates. In early 1998 the decision should be made as to who will be in the EMU and who will not. The actual start of the third stage will occur only on 1 January 1999. At that moment, the exchange rates of the member countries will be 'irrevocably' fixed both against the other member countries' currencies and against the Euro. Let us assume that this irrevocable fixing will be 100% credible. (We come back to this issue in the next section, because it is not obvious that this can be done.) Thus, 1 January 1999 will be the last time that countries can devalue or revalue their currencies. This creates a potential for great uncertainty about the intentions of the authorities and, therefore, can lead to a lot of speculative pressures during the final approach towards EMU.

Let us illustrate this problem of a 'final realignment' by taking the case of Belgium. Suppose that it is expected that the Belgian authorities will use the opportunity of the introduction of the monetary union to have a final devaluation of the Belgian franc. The temptation for Belgium to do so is strong because this final devaluation makes it possible to reduce the burden of the large government debt.[1] This opportunity is much more difficult to exploit without the monetary union. For in that case a devaluation may trigger expectations of new devaluations. As a result, the Belgian authorities may see their borrowing costs increase after such a devaluation. This will not be the case when it can be made credible that the devaluation is the last one. Only when Belgium joins the monetary union can this be achieved. After the start of EMU the Belgian authorities will not be able to change the value of their money any more. The Belgian authorities will only be able to borrow in euros.[2] Since the Belgian authorities cannot change the euro exchange rates, they can credibly promise that they will not devalue the currency in which they are borrowing. As a result, the devaluation risk premium implicit in the Belgian interest rate disappears.[3]

How can this problem be dealt with? It is important to realise that the Maastricht Treaty has introduced a feature into the decision-making process that considerably reduces the risk of such a 'final realignment'. The Treaty stipulates that the decision concerning the conversion rates (i.e. the exchange rates that will be selected for irrevocable fixing on 1 January 1999 and for final conversion into euros in 2002) must be made unanimously. This means that, in our example, Belgium will not be able to decide unilaterally about its conversion rate into euros. It will need the agreement of all members concerned. It is quite unlikely that the others will agree to such a final and significant realignment. The main reason is that such a final devaluation also improves the competitive position of Belgian industry at the expense of other participating countries, who are therefore likely to object.

Nevertheless the risk is not totally absent. Countries may come to a multilateral agreement for final realignments as part of a larger deal. Thus, although small, the risk of such final realignments is not completely zero. Which exchange rate regime can then best handle the ensuing uncertainty? There are broadly speaking two possibilities.

One regime would consist in an announcement at the moment of the decision about membership (early 1998) that the central rates (the parities) in the ERM will be the conversion rates. This can only work, of course, if the announcement can be made fully credible. If it can, the potential benefits are great, because then the market will progressively push the market exchange rates towards the central rates so that on 1 January 1999 the two coincide. The reason is that if speculators are fully confident that the central rates will be the

---

[1] The Belgian government, and for that matter many other EU governments, may have another incentive for a final devaluation: the possibility of improving the competitiveness of its industry.

[2] Note that participating countries have decided that after 1999 new government bonds will be issued in Euros only, and not domestic currencies.

[3] This does not necessarily mean that the Belgian government will be able to borrow at the same interest rate as, say, the German government. If the Belgian government borrows too much the credit rating of the Belgian government debt will be negatively affected, so that the interest rate the Belgian government has to pay will increase. We analyse this issue further in Ch. 9.

final conversion rates, they will be willing to buy a currency whose market value is below the central rate and sell a currency whose market value is above the central rate. As we come closer to the start of EMU the incentive for this arbitrage will increase so that the market rate will have to converge towards the central rate. (In box 10 we discuss a technical issue relating to the problem of whether euro-rates or bilateral conversion rates should be announced in advance.)

These benefits can only materialize if the credibility of the announcement is 100% (or very close to 100%). In order to achieve full credibility it will be necessary to set up a commitment technology at the moment of the announcement. A first element in this commitment technology can be called 'institutional front-loading'. This means that, at the moment of the announcement, some institutional changes that would normally only be implemented on 1 January 1999 should be put into practice. Thus, monetary policies of the countries accepted in EMU should already be decided upon jointly during 1998. Another part of this commitment technology consists in declaring that each participating central bank will supply its own money in unlimited amounts in exchange for the currency under pressure. At the same time the participating central banks make a commitment to target the money stock for the area as a whole. Such a solemn declaration, if credible, can beat back any amount of speculation, for the simple reason that central banks which support another currency can create unlimited amounts of their own currency to be sold in the market. Once this is known by speculators, they will not find it worth while to undertake a speculative attack. At the same time, speculators know that the targeting of the money stock of the whole area ensures that intervention activities do not affect the global money stock, since one currency is sold in exchange for another one. These interventions therefore do not affect inflation in the system as a whole.

If the participating countries are unwilling to have this commitment technology in place then it would be wise to rely on an exchange rate regime that allows for a significant degree of flexibility. This could then be organized as follows. The current normal band of fluctuation in the ERM amounts to 30%. This band seems wide enough to accommodate expectations of final realignments. That is, it is quite improbable that final realignments of more than 30% would be organized. Thus, if speculators expect that the Belgian franc will be devalued by, say, 10% this will immediately be reflected in a depreciation in the market of 10%. Interest rates in Belgium could then essentially remain unchanged, and the market rate on 31 December 1998 would become the conversion rate.

The authorities could help to stabilize such a system by announcing that the conversion rates will be the central rates. Speculators would not really believe this (because of the absence of 'institutional front-loading'), but enough uncertainty would be created in their minds so that they would find it risky to drive the exchange rate too far away from the central rate.

We conclude that if the authorities are not prepared drastically to increase their co-operation during the run-up towards the final stage of EMU, then the use of the wide band that the ERM provides will become necessary. This alternative and more flexible system can safely drive the participating currencies into EMU.

BOX 10. Announcing bilateral or euro conversion rates

In the main text we have argued that it would be desirable to announce conversion rates in advance of the start of stage three of EMU. The issue that arises here is whether one should announce the conversion rates of the participating currencies into the euro or the bilateral conversion rates (e.g. the FF/DM rate).

The Treaty of Maastricht states that the conversion on 1 January 1999 *'shall by itself not modify the external value of the ECU'* (article 109/(4)). In addition, the Madrid European Council of December 1995 decided that the substitution of the ECU by the euro on the same date shall be on a one-to-one basis. These two conditions severely limit the options available concerning the announcement of the conversion rates. We show this as follows:

Let T be the day when stage three starts, i.e. when the parities of the currencies of the in-countries are irrevocably fixed against the euro and hence bilaterally. Suppose that of the N currencies belonging to the ECU basket I are of in-countries while $N - I$ are of out-countries, not joining the single currency at the outset. We ask the following question: is it possible to announce the parities of the I currencies against the ECU (equal to one euro) at some time $t < T$, considering the two previously mentioned conditions.

From the basket definition of the ECU we know that (see box 6):

$$ECU_i = \sum_{j=1}^{I} a_j S_{ji} + \sum_{k=I+1}^{N} a_k S_{ki} \tag{B10.1}$$

Let us denote the announced fixed euro conversion rates by $Euro_i^*$. The Madrid Council decision requires that $Euro_i^* = ECU_i^*$. Triangular arbitrage ensures that, by fixing the Euro/ECU rates, the bilateral exchange rates for the in-currencies are also fixed, i.e.:

$$S_{ji}^* = \frac{ECU_i^*}{ECU_j^*} \tag{B10.2}$$

Using (B10.2) to rewrite (B10.1) and rearranging, yields

$$ECU_i^*(1 - \sum_{j=1}^{I} a_j \frac{1}{ECU_j^*}) = \sum_{k=I+1}^{N} a_k S_{ki} \tag{B10.3}$$

for each in-currency 1,. . .,I.

As can be seen from (B10.3), the left-hand side of the equation is a constant, determined by the chosen euro rates. Hence also the right-hand side, which is the weighted exchange rate of the subset of out-currencies in the ECU against each in-currency, must be a constant. In other words, pre-fixing the euro rates of the in-currencies in advance of 1 January 1999 and guaranteeing the one-to-one conversion of ECUs into euros requires that the weighted exchange rate of the subset of out-currencies with the in-currencies remains constant in the period between the announcement and the conversion. This can happen only in two cases: all the outsiders decide to fix (unilaterally) their exchange rates with the insiders at the time of announcement and are able to defend those rates; or, by some fluke, the movements of the exchange rate of the out-currencies against the in-currencies are such as to exactly offset each other. We may be allowed to say that the probability of the first case is very small and that of the second is nil for all practical purposes. Barring these two cases, pre-fixing the in-currencies' conversion rates in terms of euro violates the requirement that the external value of the ECU should not be affected by the conversion procedure. In particular it would make the dollar value of $ECU_i{}^*$ different from that of the ECU basket.

Does this problem also arise when the bilateral conversion rates are fixed in advance? One can show that if conversion rates are set as bilateral parities between the in-currencies, we restore a degree of freedom so that the external value constraint becomes less tight. Consider first the case in which conversion rates are pre-announced. Equation (B10.1) now becomes:

$$ECU_i = \sum_{j=1}^{I} a_j S_{ji}{}^* + \sum_{k=I+1}^{N} a_{ki} S_{ki} \tag{B10.4}$$

where $S_{ji}{}^*$ are the pre-set bilateral conversion rates of the in-currencies. At T the conversion rates of the latter with the euro will be the market rate of the ECU, which will depend on the bilateral exchange rates of the out-currencies. The only consequence of fixing the in-currencies' bilateral rates is that the out-currencies will appreciate or depreciate against each in-currency in exactly the same proportion.

The previous analysis assumes that the announcement of the bilateral conversion rates is fully credible so that bilateral market rates are driven to their announced conversion rates on day $T-1$. Suppose, however, that the market distrusts this announcement so that on day $T-1$ the market rates diverge from the announced conversion rates. In that case the authorities face a difficult choice which can be made explicit as follows. Writing the ECU market rate on day $T-1$ as

$$ECU_i{}^{T-1} = \sum_{j=1}^{I} a_j S_{ji}{}^{T-1} + \sum_{k=I+1}^{N} a_k S_{ki}{}^{T-1} \tag{B10.5}$$

we obtain

$$Euro_i - ECU_i{}^{T-1} = \sum_{j=1}^{I} a_j(S_{ji}{}^* - S_{ji}{}^{T-1}) +$$

$$\sum_{k=I+1}^{N} a_k(S_{ki}{}^T - S_{ki}{}^{T-1}) \tag{B10.6}$$

In order to make the one to one conversion of the ECU into the euro on day T possible, the left hand side of (B10.6) must be zero. At the start of day T the bilateral market rates of the out-currencies are equal to those observed at the end of day $T-1$. This sets the second term on the right hand side of (B10.6) equal to 0. This implies that if on day $T-1$ the bilateral market rates of the in-currencies, $S_{ji}{}^{T-1}$, diverge from the announced conversion rates, $S_{ji}{}^*$, the authorities face the following difficult choice. Either they renege their announcement and select the market rates of day $T-1$ as bilateral conversion rates, which allows them to convert one ECU into one euro on day T ($Euro_i - ECU_i^{T-1} = 0$ in (B10.6)); or they drop the latter constraint which allows them to stick to their announced conversion rates. This difficult trade-off can only be avoided if the announced fixed bilateral conversion rates are fully credible. In that case the latter will coincide with the market rates and the one-to-one conversion of the ECU into euros on day T does not pose problems.

The authorities could proceed as follows. First, the in-countries make an agreement on the structure of their bilateral exchange rates. Such agreement may or may not be announced, but it is desirable that it be reached prior to 1 January 1999. Secondly, the authorities of the in-countries take a firm commitment to steer their bilateral market exchange rates towards the agreed levels by means of coordinated intervention and interest rate policies. Whether announced or not, the markets must perceive this agreement as credible: if so, as the final date approaches, market rates will gradually converge to their target levels. Thirdly, on 1 January 1999 the euro conversion rates will be set equal to the market ECU rate of 31 December 1998. Our previous discussion shows that this can be done without affecting the agreed upon *bilateral* rates of the in-currencies.

2. How to make irrevocably fixed exchange rates credible?

To the outsider it may seem quite surprising that European leaders decided to start EMU on 1 January 1999 without a true monetary reform, i.e. without replacing the national currencies by the euro, which would then have become the single European currency with legal tender. This would have been the logical thing to do. There would have been no need to have irrevocably fixed exchange rates between national currencies and the euro for three years, a situation which can only lead to trouble. In addition, one would have avoided the awkward situation in which consumers will be able to use euro deposits, but not euro coins and notes. The latter two will only be introduced in 2002.

It has been said that this three-year transitional period, from 1999 until 2002, is necessary for technical reasons, i.e. it takes time to print the new currency, and to make the necessary changes in the national banking systems. It is difficult to take this argument seriously, however. After all, the Treaty was signed in 1991, leaving plenty of time (eight years) to prepare for the technical changes.

A more plausible explanation is that the commitment to monetary union in a number of countries was quite weak, at least initially. Thus, the additional three years of transition and the maintenance of national currencies during this period provides a last chance of opting out if 'things get out of hand'. This would not be possible if on 1 January 1999 the national currencies have disappeared.

Whatever the truth of this matter, the overriding question today is whether the transitional regime during which national currencies will circulate side by side with the euro, albeit with irrevocably fixed exchange rates, can be made trouble-free. This is the same as asking the question whether the fixity of exchange rates can be made fully credible. Fortunately, the institutional set-up that will be in place after 1999 (and which was devised by the drafters of the Maastricht Treaty) should, in principle, solve this credibility issue. Let us look at the different ingredients of this institutional structure. We can distinguish three important credibility-enhancing features.

(1) There will be one central bank, the ECB, responsible for the monetary policy of the union as a whole. This central bank, therefore, will target *union-wide* variables, e.g. the total money stock in the union. It will not care about, say, the money stock in Germany or in France. In addition, the ECB will use the euro to implement these policies, e.g. it will buy and sell euros to influence the interest rate, or to target the money stock in the union.

(2) The ESCB will be ready to convert one member currency into another member currency on demand. Thus, if say, French residents desire to get rid of their French francs and to hold euro deposits (or even German marks) the ESCB will be ready to convert these French francs into Euro deposits (or German marks). There is no limit to the conversion capacity of the ESCB, because it can 'manufacture' the euro deposits (and the German marks). Thus, speculators cannot expect that the ESCB might 'run out of money' to organize the conversion. It is precisely the expectation that the central bank will 'run out of money' that triggers speculative attacks. In addition, since the ESCB continues to target the system-wide money stock, there is no danger that this intervention will increase the total money stock in the system. The creation of German marks exactly offsets the destruction of French francs.

(3) The conversion of one member currency into the euro, or into another currency, will be at a fixed rate without a fluctuation band. This is possible because there will be no foreign exchange market in which these currencies are traded. Legally each currency will become subunits of the euro. Commercial banks will convert these currencies at the 'irrevocably fixed' exchange rate. Of course, banks will charge a fee for this service, very much in the same way as they charge fees to a customer who 'converts' a cheque into cash, or who uses his credit card. The absence of a band of fluctuation is important. The existence of such a band and the movements of exchange rates within the band has often triggered speculative pressure.

These three features are quite essential in providing the basis for the full credibility of the transitional monetary regime during 1999–2002. In order to illustrate this further we use the two-country monetary model developed in the chapter on the EMS. This will also allow us to contrast the transitional regime between 1999 and 2002 with the EMS, which suffered so much from credibility problems.

We represent the model in Fig. 7.1 and we assume that the two countries are

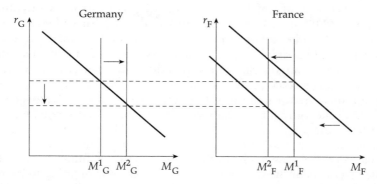

Fig. 7.1. Asymmetric shocks during the transitional period

France and Germany.[4] Let us assume that output declines in France. As a result, the money demand function in France shifts to the left because French residents now have less need to hold money for transaction purposes. The ECB is the sole monetary authority and targets the money stock of the Euro area as a whole. Initially this money stock was $M^*_E = M^1_G + M^1_F$. Thus, after the shock in France, the ECB continues to target the same money stock M^*_E. This means that a redistribution of the money stock between France and Germany will have to come about. The excess supply that now exists in France at the initial interest rate spills over into Germany. The money stock in France declines to M^2_F and the money stock in Germany increases to M^2_G. As a result, the total Euromoney stock remains unchanged. Note that the shock in France reduces the demand for money in the system as a whole. Since the ECB targets the Euromoney stock this must lead to a decline in the Euro interest rate. This, in turn, increases the demand for money in both Germany and France (we move down along the money demand functions). At the end of this process, there will be more money in Germany and less in France. Note, however, that because of the decline in the interest rate, the stock of money held in France will have declined less than if the interest rate had remained unchanged. We have seen that the latter is precisely what happened in the EMS when Germany was the leader and was targeting its own money stock. In that case the money stock in France had to decline more, exacerbating the recession, and leading to tensions between the two countries.

In practice the redistribution of money between France and Germany will be realized automatically through the banking system without the ECB having to intervene in the foreign exchange market (which will not exist between the currencies involved). This automatic redistribution of money can be described as follows. As the French economy goes through a recession, French demand for bank loans declines. French banks find themselves with excess funds, which they will invest in the interbank market. As a result, there will be excess supply

---

[4] Alternatively Germany could be considered as the rest of the union.

in that market, part of which will flow into Germany. At the same time, the excess supply in the interbank market reduces the interbank interest rate (and probably also the other short-term interest rates linked to that market). This increases the willingness of German banks to borrow funds in that market. Thus, there will be a redistribution of liquidity towards Germany, which will increase the money stock in that country.[5]

This automatic redistribution of liquidity through the banking system can only work satisfactorily if two conditions are satisfied. First, there must be full integration of the interbank markets in the system. In order to achieve this, it was agreed to set up a sophisticated payment system (called TARGET) which will make it possible for banks in the EMU to transfer funds to other banks (almost) instantaneously, very much as they are capable of doing today within their national banking systems. Second, there must be full confidence in the absolute rigidity of the exchange rate. This also implies that there should be no bands of fluctuation. For, if the latter were the case, banks would carry some risk when making transfers to another country, i.e. they would never be completely sure about the exchange rate they would have to pay when making the transfer (or the exchange rate they would receive when obtaining payment). When large shifts occur, this would limit the willingness of banks to take large open positions in the currencies involved, making it necessary that the ECB itself should intervene. The existence of a band of fluctuation would also lead to differences in the interest rates in the different currencies as the riskiness of these currencies would not be evaluated in the same way.

The previous discussion illustrates the peculiar nature of the credibility of monetary regimes. In order to have full confidence in the fixity of exchange rates during the 1999–2002 period we need the right monetary and financial institutions (e.g. full integration of the interbank market, no bands of fluctuation). These institutions, however, can only function satisfactorily if there is full confidence in the fixity of exchange rates. There will therefore be something circular in the credibility of the EMU during the period 1999–2002. Although in principle one should be able to master the potential problems, there will always remain some risk as long as national currencies remain in circulation alongside the euro.

As mentioned previously, if this risk materializes, and banks refuse to organize automatically the transfer from one money into the other, it is essential that the ESCB stand ready to supply the market with whatever currency is in excess demand and to absorb the money that is in excess supply in the private markets. Technically, the ESCB can do this without limit. In other words, the ESCB can and should perform its function of 'currency converter of last resort'. The confidence that the ESCB will perform this function will go a long way toward eliminating the risk of a speculative crisis that would destroy the fixity of exchange rates.

---

[5] Asymmetries in the size of countries matter here. If Germany is large relative to France, the increase in the German money stock will be small in relative terms.

Can one conclude that there is no risk of a collapse of the 'irrevocably fixed' exchange rate arrangement during 1999–2002? Not necessarily. The risk will continue to be a real one. This risk could materialize if the commitment to the final and crucial step of replacing the national currencies by the euro is perceived to be in doubt in one or more countries. In this connection the German Constitutional Court of Karlsruhe (Bundesverfassungsgericht) has ruled that Germany maintains its right to leave the EMU if the latter fails to provide monetary stability. Although it is difficult to interpret the meaning of monetary stability, it is not inconceivable that Germany would interpret the performance of the EMU as weak. This might then trigger suspicions that Germany would not want to make the final step of eliminating the German mark in 2002. Such an expectation would certainly destabilize the EMU during 1999–2002.

There exist some additional credibility-enhancing measures. An important one would be that the national governments issue new debt in euros from 1999 on. In fact, the prospective future EMU members have already agreed to do so. As a result, from 1999 on all *new* bonds issued by the EMU governments will be in euros rather than in the national currencies. This decision will certainly enhance the credibility of the transitional period. Some countries, e.g. Belgium and France, have gone farther and have announced that the *old* debt will also be converted into euros in 1999. Other countries, e.g. Germany, hesitate to take that step. Given the strategic position of Germany, a decision by the German government to follow the example of Belgium and France would further boost the credibility of the transitional period.

To conclude, it is worth mentioning the following paradox. In order to make the 'irrevocably' fixed exchange rate regime fully credible during 1999–2002, one needs all the institutions of a full-fledged monetary union. The European authorities decided to have all these institutions but failed to take the next logical and easy step, i.e. to introduce a *single* currency at the same time. Instead they decided to keep national currencies alive for three years, thereby reducing the credibility of the new monetary institutions. In addition, during the period 1999–2002 confusion will reign as consumers and producers face two currencies in their country. There was probably no worse way to start a monetary union. As mentioned earlier, the only reasonable explanation for having this interim period of three years has to be found in the initial lack of enthusiasm towards monetary union in some important European countries.

3. The parallel currency approach to monetary unification

The parallel currency approach to monetary unification has been hotly debated in the academic literature. The Delors Report came out against this approach, so that the issue of whether this was a useful alternative seemed to be buried. The situation that will exist during 1999–2002 when the euro (the parallel currency) will circulate alongside the different national currencies makes this approach relevant again.

The idea of using a parallel currency to force monetary integration is not a new one. Numerous academic proposals have been formulated. One of the best known is the 'All Saints' Day Manifesto' published in *The Economist* on 1 November 1975.

The basic idea of the parallel currency approach is that the creation of a European currency which is made sufficiently attractive will readily be used by economic agents, and in so doing drive the national currencies out of the market. At the end of the road we will have monetary union. One of the main intellectual attractions of this approach to monetary unification is that the speed of movement towards EMU will be determined by the free choice of European citizens, and not by decisions taken by politicians and bureaucrats.[6] The question that we want to analyse in this section is whether, and under what conditions, the euro may displace the national currencies *de facto* before the start of the final stage of EMU in 2002. If that happens, true monetary union would be forced by the market prior to the year 2002 because producers and consumers massively decide to use euros instead of the national currencies. Before analysing whether this is likely to happen it is useful to study the experience of the ECU in this matter.

3.1. The failure of the ECU as a parallel currency

The experience with the ECU suggests that the mere existence of a parallel European currency is not sufficient to trigger a dynamic towards monetary union. In fact, the original drafters of the EMS hoped that the ECU would play this role. It can now be said that the ECU has not fulfilled these expectations. The ECU has remained very marginal as a European currency. As a medium of exchange it is used very little. The great bulk of transactions in Europe are made using national currencies. Similarly, the ECU is used very little as a unit of account. Apart from the European institutions few national agents (private and official) use the ECU as a unit of account. Some economists had predicted that the ECU would be used as an invoicing currency in international trade. However, the evidence suggests that this has not happened on a large scale. Less than 1% of international trade in the EU uses the ECU.[7]

More surprising is the fact that, despite high hopes, the ECU has not really become a major currency of denomination in the international bond markets. In the early eighties there was a general expectation that the ECU could play a major role in the international bond markets. The reasoning behind this hope

---

[6] The idea has been propagated by Hayek; see e.g. Hayek (1978). Hayek added that the competition provided by this parallel currency would force the national monetary authorities to behave better. As a result, this strategy would have the interesting side-effect of reducing inflation. See also Salin (1984) and Vaubel (1978 and 1990).

[7] See Jozzo (1989) for evidence. See also De Grauwe and Peeters (1989), and Lomax (1989).

Table 7.1. The ECU in the international bond market

Currency	Stock outstanding	
	end of 1989	June 1996
	(billions of US dollars)	
ECU	54.9	67.4
DM	121.6	274.0
Yen	131.4	335.1
Pound sterling	81.6	160.7
Swiss franc	140.3	162.3
US dollars	565.9	815.6
Total (all currencies)	1,249.0	2,251.0

Source: BIS, *International Banking and Financial Market Developments*, Basle, February 1991 and August 1996.

was that the basket definition of the ECU would be attractive for the small investor wishing to hedge against exchange risk.

Table 7.1 shows that the ECU has not been able to become a major currency in the international bond markets.[8] At the end of 1989 only 4.4% of the outstanding stock of bonds issued in the international markets was denominated in ECU, a much smaller percentage than the dollar, the yen, the DM, the Swiss franc, and the pound sterling. In 1996 the share of the ECU had dropped even further to 3%.

On the whole it is fair to conclude that the ECU has not become a major currency in Europe. It is not used as a medium of exchange or as a unit of account, nor is it a major investment vehicle.

The main reason for this failure of the ECU to perform its monetary role is that it has few natural advantages. Its basket definition may be attractive to small investors; it does not, however, give the ECU an advantage when used as a medium of exchange or a unit of account. In addition, since it contains many currencies that are subject to relatively large inflation, it has a drawback for economic agents living in low-inflation countries.

Several proposals were put forward to make the ECU more attractive so that it would automatically displace the national currencies. The All Saints' Day Manifesto proposed introducing an ECU which would be inflation-proof.[9] The UK government proposed a similar plan in 1990 that would make the ECU more attractive by ensuring that it would never devalue against any of the national currencies. As a result, this 'hard' ECU would always be the strongest currency in Europe. This would also imply that the 'hard' ECU would cease to be a basket currency. It would really be a new ECU, defined in such a way that its exchange rate would remain fixed against the EC currency which appreciates most against the other currencies. According to the UK government, this

[8] It should be stressed that these numbers overstate the importance of the ECU in the bond markets. The international (Euro-)bond market is only a small fraction of the total market in bonds, which includes the national bond markets. In the latter the ECU is almost never used. See Lomax (1989) on this.   [9] For a discussion, see Fratianni and Peeters (1978).

would be sufficient to give the ECU an advantage, so that economic agents (if given the choice) would select that currency as a medium of exchange and a unit of account.

Whether such a hard ECU would have had any chance of displacing national currencies as a medium of exchange and a unit of account remains very doubtful. There is a substantial amount of empirical evidence suggesting that such an outcome is very unlikely. In a well-known study of eight hyper-inflations Cagan (1956) found that, with the exception of the German hyper-inflation of 1923 there was no significant substitution of other (low-inflation) currencies for the domestic currency. Barro (1972) estimated that the proportion of transactions made by substitute moneys (or even barter) during hyper-inflations was only 5%. (See also Klein (1978).) Latin American evidence about 'dollarization' suggests that one needs very high inflation rates before residents start substituting the dollar for the domestic currency. This empirical evidence suggests that a hard ECU would have had little chance of displacing national currencies that are subjected to at most 10% inflation a year. If history is any guide, one would have needed truly dramatically high inflations in member states of the Community to lead to significant currency substitution.

The reason why currency substitution appears to be so low has something to do with the collective good nature of money. This means that the utility an individual agent experiences from the use of a money exclusively derives from the fact that others use that same currency. This feature of money makes it very similar to a language, whose utility is determined exclusively by the fact that other individuals use it. The collective good nature of money has important implications for the substitution of one money for the other. This substitution cannot be the result of individual decisions. Individuals who have an incentive to use another money face a co-ordination problem in that many people must be willing to do the same thing. Put differently, when an individual would like to use another money he cannot decide this alone: he has to consult others, who have to agree with him to use a different money. This raises the cost of switching to another currency. (See Klein (1978) on this problem.)[10] The result of all this is that the market will not easily organize a switch to another currency when this other currency is only marginally better than the national currency. Such a switch can then only be organized by collective action. This is also the approach taken in the Maastricht Treaty.

3.2. The euro as a parallel currency

The previous ideas allow us to analyse whether the euro is likely to replace the national currencies *prior to 2002*. In other words, will the advantages of the euro be such that consumers and producers will freely substitute the euro for

---

[10] This co-ordination problem is very similar to the one that arises in the selection of standards. For example, the QWERTY keyboard is certainly inferior to other keyboards. The superiority of these other keyboards has proved insufficient to displace the QWERTY keyboard. See David (1985).

national currencies in their everyday transactions before the official switch to the euro in 2002?

Our discussion of the failure of the ECU to displace the national currencies teaches us that substitution through market forces does not occur easily, because it necessarily involves a collective decision process. This substitution must be given a push from the authorities. Otherwise it may not happen automatically. To see this, suppose governments do nothing to push the euro after 1999. Thus, citizens in each country will still have to pay taxes in their national currencies; firms will continue to have to present their accounts in national currencies; governments will still issue their debt in national currencies. If this is what happens, individual agents will have very few incentives to use the euro. Since many of their transactions and much of their accounting will have to be done in national currencies, consumers and producers will most likely continue to use these national currencies. Using a second money would not make transactions easier.

Governments, however, could stimulate the use of the euro in various ways. One has already been decided. From 1 January 1999, new government bond issues will be in euros, thereby creating a large bond market in that currency. Some countries, e.g. France and Belgium, have announced that they will convert (at least part of) the outstanding stock of government bonds into euros. Governments would strongly stimulate the use of the euro by allowing citizens to use the euro to pay their taxes and firms to use the euro for accounting purposes. Such steps will be necessary to induce agents to substitute the euro for their national currencies. It is not yet clear that governments will be willing to take these steps. The pressure to do so is increasing because many firms want to make the change-over as quickly as possible. We conclude that early switchover to the euro during 1999–2002 can occur if governments are willing to make these institutional changes prior to the year 2002 (when they are required to do so). Thus, the substitution of the euro for the national currencies will not be automatic. It will need a push from the national governments. (For an in-depth discussion of more technical problems during 1999–2002, see Gros and Lannoo (1996).)

The single most important obstacle to the widespread use of the euro during 1999–2002 comes from the fact that there will be no euro coins or banknotes in circulation during that period. (This will only happen in 2002.) As a result, consumers will have to use their national currencies for their day-to-day cash payments. Of course, they will be able to open euro accounts and make payments in euros using cheques, transfers, or credit cards. Many consumers may therefore start making most of their payments in euros. However, many others may find it unattractive to make some of their payments in national currencies (when using coins and notes) and other payments in euros. They may therefore prefer to use their national currencies. This would avoid the trouble of having to perform frequent mental exercises of converting their national currencies into euros and vice versa. Such mental exercises may be quite difficult if (as is likely) the conversion rates between the national currencies and the euro will not be round numbers.

4. How to organize relations between the 'ins' and the 'outs'?

In a previous chapter we argued that the Maastricht Treaty may lead to a split-up of the European Union into a group of countries that will participate in EMU from 1999 on (the 'ins') and another group that will be barred from entry in 1999 (the 'outs'). If that happens, the issue arises of how monetary relations between the 'ins' and the 'outs' should be organized.

There are two possible scenarios. The first scenario is one in which the membership problem and the ensuing conflict is resolved by allowing all countries close enough to (but not quite at) the Maastricht norms into the union. This would mean that practically all EU countries that have made it clear that they want to join in 1999 would be allowed into the EMU (the only possible exception being Greece). Thus, in this scenario there would be few, if any, involuntary outsiders. The problem would then boil down to organizing the monetary relations between the voluntary outsiders (the 'opting-out' countries, Great Britain and Denmark, if these countries choose to stay out) and the 'ins'.

In a second scenario the membership issue is resolved by starting EMU without countries like Spain, Italy, Portugal, and Finland. This would be a traumatic scenario since these countries' governments have invested important political capital in entering EMU. As explained in Chapter 6, the fact that they are left out would strain relations between the 'ins' and the 'outs'. In addition, these countries would only accept being left out if they received guarantees of a speedy admission to EMU. Failure to obtain such guarantees would give these countries an incentive to use their blocking minority power to vote against the admission of some of the 'ins' whose convergence is far from perfect by the standards of the Maastricht Treaty. Thus, in this scenario, the way that relations between the 'ins' and the 'outs' are organized will be very different from the first scenario.

4.1. Relations between the 'ins' and the voluntary 'outs'

The main issue that arises here is the exchange rate regime that should be set up between the euro and the currencies of the voluntary 'outs'. The United Kingdom, the most prominent voluntary outsider, has made it clear that it does not want to be constrained by an ERM-type arrangement, even the one that came into existence after 1993 and which allows for a total band of fluctuation of 30%. There is thus very little choice but to have a floating sterling–euro exchange rate. This may create problems if in the future this exchange rate fluctuates a lot and creates large movements in the competitive position of the UK economy relative to the EMU economy. It is, therefore, also unclear whether such a free-floating exchange rate arrangement can be maintained in the long run. The only conceivable constraint that could exist for the UK while it stays outside EMU is given by article 109*m* of the Treaty, which says that member states that do not participate in EMU should 'treat their exchange rate policies as a matter of common interest and it is accordingly agreed that exchange rates should be monitored and assessed at the Community level with a view, in particular, to avoiding any distortion in the single market'.

A second voluntary outsider is Denmark.[11] This country remained in the ERM after 1993 and is likely to continue to do so after 1999. As a result, very little will change except that the exchange rate arrangement of Denmark will be one of pegging to the euro instead of the German mark.

The third country which is likely to stay out voluntarily is Sweden. Since this country has not signed an opting-out clause in the Treaty, it cannot really stay out if it satisfies the Maastricht convergence criteria. However, by deciding not to join the ERM it has effectively put itself out of the running. As will be remembered, membership of the ERM is one of the convergence criteria that countries have to satisfy to be allowed into EMU.

4.2. Relations between the 'ins' and the involuntary 'outs'

The nature of the relations between these two groups of countries will be quite different from those governing the 'ins' and the voluntary 'outs'. In this case the overriding objective of these relations will be to guarantee that the involuntary 'outs' can join EMU quickly. As argued earlier, the involuntary 'outs' will not easily accept being put in the situation of outsiders unless they can obtain some guarantees that they will be able to join EMU quickly. This then leads to the question of the kind of arrangements that can speed up the convergence process and thereby ensure a quick entry into EMU.

The principles that should guide the exchange rate relationships between the 'ins' and the 'outs' were agreed upon at an ECOFIN meeting in June 1996. The main principles are the following. A new exchange rate mechanism should replace the present ERM from 1 January 1999. Adherence to the mechanism would be voluntary. (This principle was accepted, much to the chagrin of the French authorities, at the insistence of the UK government.) Its operating procedures should be determined in agreement between the ECB and the central banks of the 'outs'. The new mechanisms should be based on central rates around which margins of fluctuations should be set. The latter should be

---

[11] It should be stressed that Denmark can subject its adherence to EMU to a referendum. Therefore, we cannot rule out the possibility that Denmark will enter EMU.

relatively wide, such as the ones in the present ERM. Thus, countries may choose different margins. The anchor of the system would be the euro. When exchange rates reach the limit of the fluctuation margin, intervention would in principle be obligatory. This obligation, however, would be dropped if the interventions conflicted with the objectives of price stability in the EMU countries or in the outsider country. The ECB would have the power to initiate a procedure aimed at changing the central rates.

Many unresolved issues relating to the workings of this ERM II remain. The major risk of ERM II is that it may not be robust against strong speculative forces that are likely to arise. This has to do with the fact that the failure of the 'outs' to be accepted will be perceived as a major setback. This is then likely to intensify the 'vicious' dynamics we discussed earlier in this chapter. The 'outs' are likely to face strong downward pressures on their exchange rates, and an increase in interest rates. Increased interest rates will make the budgetary Maastricht norms more difficult to achieve. Thus, the shock of non-entry is likely to lead to renewed divergence of the 'outs' relative to the EMU countries. The interest rate effect is especially likely to lead to a setback in the convergence process of government budget deficits, forcing the 'outs' to intensify their policies of budgetary restrictions. This could in turn have severe negative political consequences as voters are likely to be fed up at the prospect of a new and painful process of less government spending and more taxes. Whether the consensus about the desirability of joining EMU can survive in these countries is very uncertain. It is not impossible that in some of these countries the desirability of the EMU project itself would be called into question. If this happens the speculative pressure exerted on the new exchange rate mechanism could become overwhelming.

In order to avoid this scenario it will be necessary to give the 'outs' sufficient guarantees that they will be able to join quickly. Our previous discussion makes clear why this is important. If the market believes that there is a large probability the 'outs' will join soon after 1999, a 'virtuous' convergence dynamics can be set in motion. In particular, the interest rates in the 'outs' will then quickly converge to the euro interest rate, thereby producing an important budgetary bonus that can quickly lead to the Maastricht budgetary numbers. How can such guarantees be given? One way is to set a date (say 2001) at which the 'outs' will be accepted. Of course, acceptance in the year 2001 cannot be made unconditionally. What can be done, however, is to stipulate that countries should continue their convergence efforts. If they can show evidence for this, nothing should stop their admittance in 2001. Such a scheme would create the virtuous dynamics that allow the 'outs' to show progress towards convergence.

Whether such a scheme is legally possible given the Maastricht Treaty (art. 109k-2) is another question. On the other hand, barring the 'outs' in 1999 while sticking to the Maastricht approach of slow convergence with uncertain entry dates will create such a setback as to make it very likely for the 'outs' to remain outside for a long period of time.

Several additional issues arise concerning the operation of ERM II. First, there is the issue of the size of the band. If this is 30% against the euro, this implies that the exchange rate of, say, the lira against the peseta (if these two currencies are outside EMU) can fluctuate up to 60%.[12] One could therefore choose a smaller band of fluctuation against the euro (e.g. 15%).

Second, there is the issue of the intervention commitment of the ECB. The principles agreed upon at the moment stipulate that the ECB should stand ready for unlimited intervention when, say, the lira drops below the lower limit of the band, unless this intervention would jeopardize price stability in the EMU. This may seem to seriously reduce the intervention commitment of the ECB. However, it is very unlikely that interventions by the ECB would do this. First of all, the outsider countries are likely to be small relative to the whole of the EMU area, so that the intervention activities of the ECB will have a small quantitative impact in the EMU money market. In addition, the ECB is likely to sterilize these interventions so that they will not affect the EMU money markets. It is therefore reasonable to conclude that if the ECB is committed to intervening, it can do so without jeopardizing price stability in the EMU.

To conclude, it should be stressed that such an ERM II arrangement only makes sense in the framework of a temporary regime that facilitates the quick convergence and acceptance of the 'outs' into EMU. If the 'ins' are unwilling to give a guarantee for rapid entry of the 'outs' then an ERM II arrangement may be undesirable. It may then quickly face problems similar to those the EMS experienced in 1992–3, with speculative crises and a collapse of the arrangement.

---

[12] The reason is that if, say, the peseta is at its upper limit against the euro and the lira at its lower limit, a drop of the peseta against the euro by 30% and a simultaneous increase of the lira by 30% against the euro would imply a 60% depreciation of the peseta against the lira.

8

THE DESIGN OF A EUROPEAN CENTRAL BANK

Introduction

In Chapter 6 we derived the proposition that the incentive for a low-inflation country to join a monetary union with high-inflation countries is low. There is nothing to be gained for the low-inflation country in terms of inflation and unemployment. All the gains are for the high-inflation countries. In this chapter we analyse the implications of this asymmetry for the design of the European Central Bank (ECB). We also study some of the operational problems the ECB will face.

1. The design of the European Central Bank

The strong asymmetry in incentives for high- and low-inflation countries to enter EMU is very important in the design of the European Central Bank. It means that there is only one way for the low-inflation country to agree to move to the final stage. The condition is that the European Central Bank is as 'hard-nosed' about inflation as the low-inflation country's own central bank. Failure to devise such an institutional copy of the low-inflation central bank leads to a situation where the low-inflation country refuses to make the final step to EMU.

There is one way in which the incentives for the German monetary authorities to join EMU can be improved. That is to ensure that the individuals who will run the ECB are even more 'hard-nosed' about inflation than the German authorities themselves. Put differently, if the German monetary authorities are forced to accept EMU, they will probably insist on having an ECB that gives an even higher weight to price stability than the Bundesbank does today.

The striking fact is that the Maastricht Treaty has gone a long way in this direction. The statutes of the ECB give even more emphasis to price stability than do the Bundesbank statutes. In devising these statutes the Treaty has implemented two important principles (which also form the basis of the Bundesbank statutes). The first principle is that the primary objective of the European Central Bank should be the maintenance of price stability (see article 105).

The second principle is political independence. This second principle is seen as a necessary condition to ensure that the budget deficits of the national and European governments will not be financed by printing money. In an institutional environment where the central bank is an appendix of the ministry of finance (as is the case in many countries in the world) it is inevitable that the central bank will be forced to finance the budget deficits. This is the surest way to produce inflation. Thus, it can be said that this principle of political independence is the necessary condition for making the achievement of the first principle (price stability) possible.

It is useful to quote the Treaty explicitly to realize how carefully it has formulated this principle of political independence:

When exercising the powers and carrying out the tasks and duties conferred upon them by this Treaty [. . .] neither the ECB nor a national central bank, nor any member of their decision-making bodies shall seek or take instructions from Community institutions or bodies, from any Government of a Member State or from any other body. (article 107)

In addition, the Treaty has the following sentence:

Overdraft facilities or any other type of credit facility with the European Central Bank or with the national central banks of the Member States [. . .], with Community institutions or bodies, central governments, regional or local authorities, public authorities [. . .] shall be prohibited, as shall the purchase directly from them by the ECB or national central banks of debt instruments. (article 104(1))

This seems to confirm that the drafters of the statutes of the ECB have understood the basic asymmetry in the incentives for countries to join the EMU. As a result, they have taken pains to ensure that the ECB, at least on paper, will be an institution akin to the Bundesbank. In fact, the language used by the drafters of the statutes of the ECB is tougher on inflation and political independence than the statutes of the Bundesbank. The political independence of the ECB will probably be greater than that of the Bundesbank. The reason is that a simple majority in the German parliament can change the statutes of the Bundesbank if German politicians are dissatisfied with the record of the Bundesbank. Changes in the statutes of the ECB are much more difficult. Such changes can only occur by a revision of the Maastricht Treaty, requiring unanimity. (This feature has in fact led some to criticize the ECB for absence of democratic accountability. We will come back to this issue.)

The fact that the ECB has incorporated the two principles that form the basis of the Bundesbank statutes may have convinced Germany to join EMU; it does not necessarily mean that these principles are desirable. There is, however, a large amount of theoretical analysis and empirical evidence that has convinced many economists that these principles are desirable.

2. Political independence and inflation

The intellectual case for political independence of the central bank has been much influenced by the Barro–Gordon model. As we have seen, this model predicts that monetary authorities who try to improve the inflation–unemployment trade-off by the systematic use of inflationary policies will just create an inflation bias without obtaining a permanently lower rate of unemployment. Because of the short time horizon of the political decision-making process the danger exists that politically influenced central bankers will have greater incentives to produce surprise inflation so as to gain politically from temporarily lower unemployment. This has led to the idea that the conduct of monetary policies should be delegated to an independent authority which is less subject to these incentives.[1]

This idea that political independence will lead to less inflation has been subjected to much empirical analysis recently. There have been studies by Bade and Parkin (1978), Demopoulos *et al.* (1987), and more recently by Grilli *et al.* (1991), Cukierman (1992), Alesina and Summers (1993), and Eijffinger and Schaling (1995) showing that central banks that are politically independent tend to produce less inflation than central banks that have to take orders from the government. In Fig. 8.1 we show an example of such an empirical test. On the vertical axis the average yearly rate of inflation of industrial countries (1972–91) is represented. On the horizontal axis an index of political independence of the central banks as computed by Cukierman (1992) is represented. We observe that there is a negative relationship, i.e. countries where the central banks have a great deal of political independence enjoy a lower rate of inflation, on average.[2]

Of course, many problems arise in testing this hypothesis. One concerns the measurement of political independence. It must be said that much careful analysis has been performed using alternative measures of political independence. On the whole the empirical results remain robust for these alternative measures.

An important aspect of these empirical studies is that they also reveal that on

[1] See Rogoff's well-known paper (Rogoff 1985*a*). For a recent formulation see Walsh (1995). There are certainly problems with this view. Ultimately, politicians who delegate their power to conduct monetary policies to an independent authority can always take away this power. It remains true, however, that such delegation contracts can have a useful function of restraining politicians because it is not always easy to abrogate these contracts.

[2] A regression exercise involving 21 industrial countries confirms this conclusion:

$$\text{inflation} = 11.2 - 9.4 \text{ independence} \qquad R^2 = 0.88$$
$$\qquad\quad (8.7) \quad (-2.9)$$

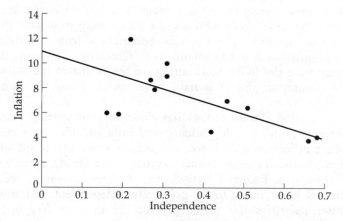

Fig. 8.1. Average annual inflation and political independence (1972–1991)
Source: de Haan and Eijffinger (1994).

average and in the long run political independence does not lead to more unemployment or to a lower growth rate of the economy. All this seems to suggest that the basic Barro–Gordon paradigm is the correct way to view reality, and that political independence is a desirable institutional feature of the future ECB.

3. The design of the ECB: Further issues

Whereas there is a broad consensus today on the need for political independence of the ECB there are still a lot of unresolved issues concerning the detail of the design of the ECB and concerning the conduct of monetary policies. In this section we discuss some of these issues.

3.1. Is political independence sufficient to guarantee price stability?

The question that arises here is whether the explicit recognition in the statutes of the ECB of political independence and of price stability as the primary objective of monetary policy is sufficient to guarantee an inflation-proof ECB. One can express doubts about this. Certainly, German politicians have doubts, otherwise they would not insist so much on the convergence criteria as additional guarantees for price stability in the future EMU and on a stability pact.

There are good reasons to believe that the actual practice of monetary policy-making may deviate from the principles embodied in the Treaty. The reason is that the individuals who will be conducting monetary policy are subject to social and cultural influences. Some come from countries where abhorrence of inflation is not as intense as in Germany. They may, therefore, act differently than the individuals sitting on the board of the Bundesbank, even if the statutes of the ECB have been copied from the Bundesbank statutes.

In this connection, Posen (1994) has done very interesting research on the link between political independence and inflation. His main conclusion is that both political independence and inflation are the result of deeper social and economic interests. Some countries have strong pressure groups against inflation (e.g. financial institutions). In these countries we observe that the central bank tends to be politically independent and inflation is low. In other countries the major pressure groups are less opposed to inflation. In these countries central banks will be less independent and inflation will be higher. This research teaches us that central banks' behaviour is very much influenced by the underlying social and economic forces, so that a mere change of the statutes of the central bank will not by itself change behaviour.

One should not, however, go too far towards a neo-Marxist interpretation of the issue. In this interpretation economic forces drive the institutions. A more balanced view recognizes that institutions (and incentives) can also change behaviour. Thus, the incorporation in the statutes of the central banks of political independence as a means of guaranteeing price stability can help to influence behaviour and to change society's view about the role of monetary policy. In addition, there is scope for strengthening these institutions so that the risk of inflation is reduced.

In the previous chapter we argued that the convergence criteria *per se* do not give Germany guarantees that the future EMU will produce price stability. We also argued that better guarantees can be given by strengthening the ECB so that it has the right incentives to produce low inflation. How can this be done? Several proposals have been made in this connection.

A first institutional strengthening consists in defining and enforcing a procedure for removal of the board of directors of the ECB should it fail to maintain price stability. In this connection, alternative incentive schemes have been proposed whereby the salaries of the ECB policy-makers would be reduced if the inflation rate exceeds some level.[3] Such incentive schemes would do more to ensure price stability in the union in, say, the year 2010 than the insistence that countries reduce their inflation rates and their

---

[3] See e.g. Neumann (1990) and Vaubel (1989). Recently, Walsh (1995) has formalized this idea in the context of a Barro–Gordon model. We analyse this model in Box 11. It should also be mentioned that New Zealand has introduced penalties to its central bank law: failure to hit the inflation target can lead to the removal of the governor of the central bank.

budget deficits in the second half of the 1990s, before the union starts. Such a reform also goes some way toward making the future European Central Bank more accountable. (We return to this issue of accountability in Section 3.3.)

Secondly, the budgetary process in the different EMU countries should be reformed so as to make it less prone to unsustainable budget deficits. This can be done in two ways. One is by implementing rules of mutual control over the budgetary processes of each participating country. This is what the 'stability pact' that was agreed upon by the EU countries at the Summit Meeting of December 1996 in Dublin aims to achieve. We will analyse this stability pact in greater detail in the next chapter. A second (and complementary) approach has recently been suggested by Eichengreen and von Hagen (1995). They formulate proposals aimed at making the budgetary process more streamlined in the European Union. In addition, they propose instituting National Debt Boards in each country, whose responsibility it would be to monitor the evolution of the national debt and to propose remedial action when particular targets are not met.

3.2. Which inflation target?

The social acceptability of the future policies of the European Central Bank will be very much influenced by the inflation target the ECB is pursuing and by its willingness to compromise when economic activity is negatively affected. Thus, two issues arise. One is the level of inflation the ECB should aim at; the other is the variability it should allow around the targeted level.

The optimal inflation rate

In the 1950s Milton Friedman (1969 edn.) formulated the view that the optimal inflation rate is zero. The basic reason for this conclusion is that a zero inflation rate maximizes the total utility of holding money. Two other factors, which recently have been much researched, cast doubts on this conclusion.

First, there is evidence that the conventional measures of inflation (the rate of change of the consumer price index) tend to overestimate the true inflation rate by 1 to 2% a year (see Gordon (1996), and Shapiro and Wilcox (1996)). One of the reasons is that the conventional measures of inflation do not take into account quality improvements.[4] As an example, take a personal computer

---

[4] There are other reasons too. For example, when the price of a particular commodity increases, consumers will substitute a cheaper one. Measurements of inflation typically disregard these substitution effects, because the consumer price index keeps the weights of the different commodities constant (most of the time).

in 1997 and compare it to one in 1980. Their price may be approximately the same. However, the computing power of the 1997 version is probably a hundred times greater, if not more. As a result, the price per unit of computing power in 1997 is a very small fraction of what it was in 1980. Many similar examples can be given. We conclude that if we observe an inflation rate of 1 to 2%, the true underlying inflation rate is probably zero.

Second, there are theoretical arguments to be made for a rate of inflation a little higher than 0%. The main one is that sectoral or micro-economic shocks require adjustments in relative real wages. In particular, sometimes a sector or a firm is confronted by a negative shock necessitating a decline in the real wage level. If the rate of inflation is zero, such a decline in the real wage can only come about by a decline in the nominal wage rate. If, however, inflation is positive one can achieve a decline in the real wage by keeping the nominal wage increases below the rate of inflation. There is a lot of evidence that resistance to nominal wage reductions is high, thereby limiting real wage adjustments when the rate of inflation is zero. Put differently, in an environment of zero inflation, there is likely to be more real wage rigidity, making adjustments to asymmetric sectoral shocks more difficult to achieve. In a recent article Akerlof *et al.* (1996) come to the conclusion that this effect may require the monetary authorities to target an inflation rate of 1 to 2% per year.

The previous analysis then leads to the conclusion that the optimal inflation rate may be of the order of 2 to 4% per year (1 to 2% on account of the measurement bias, and 1 to 2% on account of the real wage effect).

The policies pursued by the European national central banks in the 1990s suggest that they target an inflation rate below 2 to 4% per year. The central banks that are generally hailed for their anti-inflationary success have lowered inflation below 2% per year. There is a great likelihood that the ECB will also want to keep the annual inflation rate below 2% per year. Thus, it appears that the ECB will be pursuing a target inflation rate which may be too low compared to the optimal one, thereby increasing the risk of protracted deflationary monetary policies and a reduction of real wage flexibility.

The previous discussion raises the issue of who should determine the inflation target. The Treaty has not really settled this issue. It only stipulates that the ECB should pursue price stability. Given the lack of precision in this formulation it is likely that the ECB will be quite autonomous in deciding what this means. If there is a significant difference between the target pursued by the ECB and the one which society finds optimal the legitimacy of the ECB may be called into question. We return to this issue when we analyse the accountability of the ECB.

Short-run fluctuations in inflation

A second problem that arises has to do with the degree of compromise the ECB will accept for short-term fluctuations in the inflation rate around the targeted one. In this connection the Treaty stipulates the following:

Without prejudice to the objective of price stability, the ECB shall support the general economic policies in the Community with a view to contributing to the achievement of the Community as laid down in article 2. (article 105(1); article 2 of the Treaty defines these objectives, which include 'a high level of employment')

Thus, the Treaty recognizes the need for the ECB to pursue other objectives. One may debate whether the ECB should do this. The point, however, is that most societies expect the central bank not to abandon completely the ambition of stabilizing the economy. With each recession, social pressure will accumulate, pushing the ECB to relax its monetary policy stance. The problem that arises here is that the ECB will determine in a sovereign way whether and to what extent it is willing to allow for short-term deviations of inflation from its target level and whether it is willing to accommodate its monetary policies to a recession. This raises the issue of the accountability of the ECB.

3.3. The accountability of the ECB

Whereas the Treaty is quite explicit in formulating the principle of political independence, it has little to say on the issue of accountability.[5] This issue can be formulated as follows. When society delegates the task of running monetary policy to an independent institution, this delegation should really be interpreted as a contract. In this contract the objectives to be pursued by the central banks are formulated (e.g. price stability, however defined). In addition, society may wish to make the central bank politically independent because it feels that this is the best way to achieve the ultimate objectives of monetary policy. Any contract, however, implies that there must be some procedure by which society evaluates the performance of the central bank in achieving the chosen objectives (i.e. the objectives chosen by society). And here it must be said that the Treaty has created an institution with weak accountability. In article 109*b* of the Treaty it is said that 'the ECB must address an annual report on its activities to the European Parliament, the Council and the Commission. [. . .] The President of the ECB shall present this report to the Council and the European Parliament, which may hold a general debate on that basis'. In addition, the President of the Council and a member of the Commission may participate, without having the right to vote, in meetings of the Governing Council of the ECB (article 109*b*–1).

The problem of accountability arises at different levels. First, society must

---

[5] For a discussion of independence and accountability of central banks see Roll (1993).

choose the objectives for the central bank (and not the other way around). As argued earlier, the Treaty is so vague about objectives that it is quite likely that the ECB will in fact fix them. This may lead to tension when the ECB pursues objectives that are not shared by the rest of society. We mentioned the possibility that the ECB may choose a target level of inflation that is too low for the welfare of society. Or, the ECB may accord too little importance to short-term output stabilization (too little in comparison with the expectations about this in the rest of society). In all these cases, tensions may lead to a feeling that the ECB lacks legitimacy, which may endanger the long-term survival of the monetary union.

A second problem has to do with accountability in the narrow sense. Once the objectives have been determined, there should be a procedure that allows a society to evaluate how well these objectives have been achieved. If there is a perception of systematic failure a sanctioning mechanism should exist that allows society to exert sufficient pressure to redress the situation.

It must be said that the Treaty fails to set up such a procedure of accountability. It is interesting to note here that the Bundesbank, which served as the role model for the ECB, can be said to be more accountable than the ECB. The reason is that the law which describes the responsibilities and duties of the Bundesbank can be changed by a simple majority in the German parliament. This is a very direct way in which society can pressure the Bundesbank to pursue society's interests. This is not the case with the ECB, the statutes of which are part of the Maastricht Treaty, and therefore can only be changed by unanimity. There is very little society will be able to do to change the behaviour of the ECB if that institution fails to achieve the objectives society has entrusted it to pursue.

In a previous section we argued that a sanctioning mechanism should be set up. We proposed that this should take the form of a procedure to remove the board of directors should they systematically fail to achieve particular objectives. Other mechanisms can be developed. The important thing is that they should exist.

It is clear that a balance must be struck between political independence and accountability. This can be achieved by making the contract that exists between the ECB and the rest of society more explicit than is the case now. If this is done, the explicit recognition of political independence becomes part of the contract. Political independence and accountability can be made consistent. In Box 11 we elaborate on this theme using a formal model.

Box 11. The ECB: Independence and accountability

In this box we analyse issues of the political independence and accountability of the ECB using the Barro–Gordon model and recent theoretical discussions of these issues. Suppose society has a particular target for inflation and for unemployment. In addition it wants to minimize deviations from these targets. We can represent these preferences by the following loss function:

$$L = (\pi - \pi^\star)^2 + b(u - u^\star)^2 \qquad (B11.1)$$

where π is the rate of inflation and $\pi^\star$ the rate of inflation desired by society; u is the unemployment rate and $u^\star$ is the desired one; b is the weight given to the stabilization of the unemployment rate around the target level.

This loss function can be given a geometric interpretation as indifference curves around the targets $\pi^\star$ and $u^\star$, as we did in previous chapters. Society cannot run monetary policies itself. It must delegate the job to an institution, the central bank. Let us assume it appoints a central bank with the same preferences as given by equation (B11.1). Using the Phillips curve analysis, as we did in previous chapters, we can find the rational expectations solution as the point on the vertical Phillips curve where a short-term Phillips curve is tangent to an indifference curve of the central bank. We represent this in Fig. B11.1. (Note that we have normalized the targets $\pi^\star$ and $u^\star$ so that they coincide with the origin. This does not mean that the targets are equal to zero. We only do this for graphical convenience.)

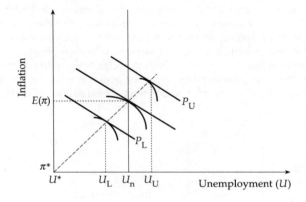

Fig. B11.1. Inflation bias and stabilization

We know from our previous discussion of the Barro–Gordon model that the greater the weight given to unemployment stabilization, b, the higher the inflation bias. Graphically, a high b implies a steep 'expansion path' (the dotted line through the origin in Fig. B11.1).

In Fig. B11.1 we assume that there are unpredictable disturbances in output and employment so that the short-term Phillips curve shifts up and down. We show this by the curves P_L and P_U, which represent the range between which the

short-term Phillips curves can shift. Given the authorities (and society's) preferences, as represented by the 'expansion path', they will aim for the unemployment points given by U_L and U_U.

As was discussed earlier, this stabilization effort is also the source of the
inflation bias, which is equal to the distance between the average inflation
rate, $E(\pi)$, and the target inflation rate, π^*.

How can society improve this situation so that average inflation comes closer
to the target rate of inflation? One possibility has been suggested by Rogoff
(1985a) and is also implicit in the Maastricht Treaty. This consists in appointing an
independent and 'conservative' central banker. Conservative here means some-
one who is willing to attach a lower weight than society to the stabilization effort.
Thus, we can represent the loss function of this conservative central banker by

$$L_c = (\pi - \pi^*)^2 + b_C(u - u^*)^2 \qquad \text{(B11.2)}$$

where b_C is the weight given by the conservative central banker to the stabilization effort, and $0 < b_C < b$.

We now contrast the solution obtained by the conservative central banker with
the previous solution, where the central bank faithfully reflects society's preferences in Fig. B11.2. The conservative central banker operates along the flatter
expansion path. As a result the inflation rate will on average be lower than when
the central banker carefully applies society's preferences. Thus, it appears that
society gains from appointing this conservative central banker. This is not necessarily true, however. For the conservative central banker applies less stabilization
effort than society desires. Thus, unemployment will fluctuate more. (We have
assumed the same horizontal displacements of the short-term Phillips curves in
the two regimes.) For example, if there is a recession, the conservative central
banker will use less stimulus, so that unemployment increases more than in the
case where the central bank pursues the same stabilization objectives as society.

This situation may lead to a problem with the legitimacy of the central banker if
society thinks that the gain in terms of lower average inflation does not compensate for the loss in terms of a suboptimal stabilization of random shocks.

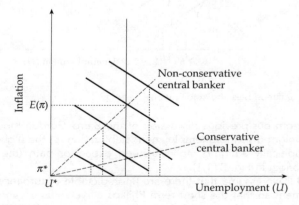

Fig. B11.2. A conservative central banker

Another problem with the conservative central banker is that there is no procedure for evaluating his performance. Put differently, the conservative central banker is not accountable.

How can this problem be solved? Recently, Walsh (1995) has shown that it can be solved by designing a performance contract with the central banker in which the latter pays a penalty if inflation is too high. Walsh has shown that a simple linear penalty rule will do the trick. We now have the following loss function:

$$L = (\pi - \pi^*)^2 + b(u - u^*)^2 + c(\pi - \pi^*) \tag{B11.3}$$

where we have added the term $c(\pi - \pi^*)$. This expresses the penalty if inflation exceeds the target inflation rate π^*. The parameter c represents the amount of dollars deducted from the central banker's salary for every point that inflation exceeds the target level. An alternative interpretation of the term $c(\pi - \pi^*)$ is the probability of being sacked, whereby this probability is a linear function of the deviation of inflation from its target level.

We show the solution in Fig. B11.3. The effect of the linear penalty rule is to shift downwards the expansion path along which the central banker operates. The intuition is that this penalty scheme has the same effect as reducing the inflation target. We can now find an optimal penalty. This will be obtained when the downward shift of the expansion path is such that a new equilibrium is reached at point U_N, where the target inflation rate is reached. In other words the penalty must be high enough so that the central banker is induced to set a target inflation rate that will exactly offset the inflation bias $(E(\pi) - \pi^*)$.

The nice thing about this solution is that the central banker pursues exactly the same stabilization effort as society desires. This can be seen graphically from the fact that the central banker operates along an expansion path which has the same slope as society's. Thus, the long-term legitimacy of this central banker is

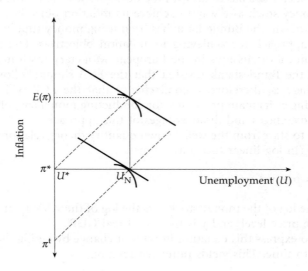

Fig. B11.3. An accountable central banker

likely to be stronger than in the case of the conservative central banker. In addition, this solution combines independence with accountability. The central banker has a contract with society to achieve certain objectives and is evaluated ex post on his performance. Apart from this, he is independent. This independence has also been called instrument independence: given the central bank's target, as it is fixed by society, the central bank is free to use the instruments needed to achieve the target.

A performance contract as proposed by Walsh (1995) is not the only way to eliminate the inflation bias while maintaining society's desire for stabilization. Recently Svensson (1995) has shown that inflation targeting can be made equivalent to Walsh's linear penalty rule. This can easily be seen in Fig. B11.3. Society now makes a contract with the central bank in which it sets the inflation target at the level π^t. The central bank pursues this target and will on average achieve π^* (which is the socially desired rate of inflation). In this inflation target contract the central bank attaches the same weight to stabilization as society. Thus, in this contract we do not need to impose a penalty on non-performance. The problem with this kind of contract, however, has to do with its lack of credibility. As can be seen, the central bank announces a target π^t which it misses systematically. This may in the end undermine the credibility of the contract.

3.4 How to conduct monetary policies?

In the previous sections we discussed the fundamental issues relating to the design of the ECB. Once these issues are solved, other issues of a technical nature, relating to the conduct of monetary policies by the future ECB, must be tackled. One such issue has to do with the question of whether the ECB should target the money stock as a way to achieve its inflation objective.

As is well known, the Bundesbank has been using money supply targeting as its favoured approach for achieving its inflation objectives. During the preliminary technical discussions in the European Monetary Institute, the representatives of the Bundesbank insisted that the ECB should follow the same approach. There is, therefore, a good chance that the ECB will use money supply targeting as its main procedure for conducting monetary policies. Let us analyse the advantages and disadvantages of this approach.

It is useful to start from the well-known quantity theory relationship, which we can write (in log-linear form) as follows:

$$m + v = p + y \tag{8.1}$$

where m is the log of the money stock, v is the log of the velocity of money, p is the log of the price level, and y is the log of real GDP.

We can also express this equation in rates of change by taking the derivative with respect to time. This yields (after rearranging)

$$\dot{m} = \dot{p} + \dot{y} - \dot{v} \tag{8.2}$$

Money targeting now implies that the authorities fix the growth rate of the money stock, $\dot{m}$, so as to achieve a particular inflation target $\dot{p}$. In order to do so, the authorities must make a forecast of the future growth rate of output $\dot{y}$ and velocity $\dot{v}$. Suppose they forecast output to increase by 2% a year and velocity to remain constant. Then equation (8.2) tells us that in order to achieve an inflation target of, say, 2% the money stock should grow at 4% per year.

This framework became popular at the end of the 1970s, when some major central banks (e.g. the US Fed and the Bank of England) switched to money supply targeting. Since then, enthusiasm about this way of setting monetary policies has waned. The reasons are the following.

First, the concept of money stock is very elusive. Should one use M1, M2, or a broader concept of money stock? Quite often these different concepts of money stock have moved in opposite directions, giving very different signals to the monetary authorities.

Second, money stock figures are released with a delay of one to several weeks and are often of poor quality (so that they have to be revised later). This problem does not exist with interest rates, which are known almost instantaneously and are more reliable.

Third, and more importantly, the money stock is an intermediate target (inflation being the ultimate target). As equation (8.2) makes clear, the precision with which targeting the money stock will bring us close to the ultimate target depends on the precision with which output growth and velocity growth can be forecast. Major problems have arisen with forecasting velocity growth. This is due to the rapid rate of financial innovation, which has led to much unpredictable behaviour of velocity.

The result of all this is that most central banks that have attempted to apply money supply targeting have been quite unsuccessful, and have missed their announced targets often and by wide margins. As a result, these central banks have returned to a more eclectic approach in which other intermediate targets like the interest rate play a role together with the money supply.

The only major central bank which continues to use money supply targeting is the Bundesbank. The evidence is certainly not that the Bundesbank has been very successful in hitting its money supply target. On the contrary it has missed the target by wide margins in many cases. The surprising thing is that this has not reduced its reputation. The reason may be that this failure to hit its money supply target has not prevented the Bundesbank from achieving its inflation target.

Should the ECB copy the Bundesbank and use money supply targeting as its main operational procedure? An argument in favour of doing this is the following. Since the Bundesbank has been so successful in keeping inflation low and since it has done this within the framework of money supply targeting, the ECB will gain from the reputation that this combination has produced by doing exactly the same thing.

The arguments against this view are quite strong though. First, it is not clear that the ECB can afford to miss the target most of the time. This is likely to happen if we can extrapolate from the experience of many central banks that have used this procedure. The Bundesbank, with its strong reputation as an inflation fighter, may get away with making large errors in reaching its money stock target. It is unclear that the ECB, at least initially, will get an equally benevolent treatment from the market.

Second, the monetary institutions in the EMU will be in great flux initially. This is likely to lead to large and unpredictable shifts between the different concepts of the money stock. It will also lead to large fluctuations in velocity. All this suggests that it would be quite unwise to use money supply targeting as the main operating procedure. This can also be seen from equation (8.2), whereby it should be remembered that $\dot{v}$ is a function of the interest rate (when the interest rate increases velocity increases, and vice versa). Let us therefore rewrite (8.2) as follows:

$$\dot{m} = \dot{p} + \dot{y} - \dot{v}(\dot{r}, EMU) \tag{8.3}$$

where velocity is a function of the interest rate, $\dot{r}$; and the variable, called *EMU*, expresses the money market disturbance that will arise at the start of EMU.

During the early phase of the EMU, velocity will be subjected to large and unpredictable disturbances. If $\dot{m}$ is fixed then these exogenous velocity shocks will be offset by interest rate movements in the other direction (assuming that in the short run $\dot{p}$ and $\dot{y}$ are sticky). The result will be that during the initial phases of EMU, money supply targeting may lead to a large variability in the EMU interest rates which is likely to influence aggregate demand and therefore output and employment. Hardly a promising start.[6]

It may, therefore, be better, at least initially, to take an eclectic view and to use both the money stock and the interest rate as intermediate targets to achieve a given inflation target. In this connection the euro–dollar rate could also be used as an intermediate target to achieve a particular target for inflation.

3.5 Monetary policy and the exchange rate

A final issue we should discuss has to do with the link between the monetary policies of the ECB and the exchange rate of the euro *vis-à-vis* outside currencies.

The Treaty stipulates that, although the ECB will independently decide the conduct of monetary policies, it will not independently decide the euro

---

[6] The same result can be shown using the IS–LM framework which was presented in Box 4 in Ch. 3. During the early phase of EMU the shocks in the LM curve are likely to dominate the shocks in the IS curve. As a result, pegging the interest rate is likely to lead to less variability in the output level.

exchange rate with non-EU currencies. In fact in article 109, the Treaty states that the Council of Ministers (i.e. politicians) shall decide whether the euro can enter into formal exchange rate arrangements with third currencies, and if so, it will be the Council who decide to devalue the euro. In the absence of a formal agreement (which is the most likely future situation) the 'Council may formulate general orientations for exchange rate policy in relation to these currencies' (art. 109(2). The Treaty adds, however, that these general orientations 'shall be without prejudice to the primary objective of the ECB to maintain price stability'.

Some observers have argued that the power of the Council to determine the exchange rate policies in relation to third currencies undermines the political independence of the ECB and may lead to conflicts about desirable monetary policies.

It is clear that if the Council decided to conclude a formal exchange rate arrangement with, say, the dollar, this would severely constrain the ECB's ability to pursue independent monetary policies. For we know that if the exchange rate is fixed, the monetary authorities lose their power to influence the domestic money stock and the domestic interest rate (at least in a situation of free capital mobility). In a fixed exchange rate environment between the euro and, say, the dollar, there would be very little scope for the ECB to fix a target for the EMU money stock (or the interest rate). As a result, targets for the inflation rate would also have to be abandoned. Thus, if the Council were to exercise its power to fix the euro exchange rate with, say, the dollar, the political independence of the ECB would exist only on paper. One can therefore only hope that the Council will not exercise this power.

The chances are good that the Council will not try to fix the euro exchange rate with outside currencies, and that the exchange rate regime with third currencies will be flexible (as it has been for most of the last 25 years). In such an exchange rate regime, there is relatively little the Council will be able to do. True, according to the Treaty it can 'formulate general orientations for exchange rate policy'. However, in the event that the Council recommends a particular exchange rate policy that conflicts with the inflation target of the ECB, the latter will be able to claim that according to the Treaty its main responsibility is price stability, so that foreign exchange market interventions should not interfere with its monetary policy stance. In addition, the ECB is likely to argue that its power to influence the exchange rate is limited, at least if it sticks to its inflation target.[7]

In this connection two additional points should be made. First, the previous discussion does not mean that conflicts will not occasionally arise between the Council of Ministers and the ECB. Some governments, e.g. the French government, may push for a weaker euro relative to the dollar so as to improve

---

[7] Empirical research tends to confirm this. The experience with interventions in the major foreign exchange markets is that they are quite ineffective if domestic monetary policies are unchanged (see Dominguez and Frankel (1993)).

European competitiveness. If the ECB fears that a depreciation of the euro would jeopardize its inflation objective, a conflict between the Council and the ECB would emerge. It is difficult to see how this would be resolved without one of the two opponents having to backtrack.

Second, as mentioned earlier, the ECB could also use the euro exchange rate as an intermediate target in trying to achieve its inflation target. Thus, when for example the euro appreciates *vis-à-vis* the dollar the ECB could compute the effect of this appreciation of the euro on inflation. In general, this would tend to reduce inflation, so that the appreciation of the euro could be a signal that monetary policies can be relaxed.

4. Conclusion

In this chapter we analysed issues relating to the design of the future European Central Bank. We argued that the political independence granted to that institution is quite important in creating the condition for price stability. We also argued that although necessary, political independence is not sufficient to guarantee price stability. We therefore formulated proposals to strengthen the European Central Bank, thereby improving the incentives for Germany to accept the creation of an EMU which also includes EU countries with a history of high inflation. These proposals also make it possible to avoid a long-term split in the European Union.

A major problem in the design of the European Central Bank concerns the balance between independence and accountability. It must be said that, while the Maastricht Treaty is very explicit in guaranteeing political independence, it is much less so on the need for an accountable central bank. This is quite unfortunate. For it is essential that the central bank should be held accountable for failures in its monetary policies. After all, the individuals managing the central bank may be motivated by objectives that do not coincide with society's interest. Like any other government institution the ECB must, therefore, be subject to outside evaluation to see whether it achieves the objectives society has contracted with it to pursue. We argued that a control mechanism should be instituted to avoid the ECB losing its legitimacy. We also argued that this can be done without compromising on the independence of the ECB.

Finally, we studied technical issues of monetary control by the future ECB. Our main conclusion here is that, at least initially, the ECB should not use money supply targeting as its sole procedure of control. Instead, it should take an eclectic stance, allowing both the money stock and the interest rate to be used as intermediate targets for achieving price stability.

9

FISCAL POLICIES IN MONETARY UNIONS

Introduction

The traditional theory of optimum currency areas, which was discussed in Chapter 1, offers interesting insights about the conduct of national fiscal policies in a monetary union. In this chapter we start out by developing these ideas. We then challenge this theory by introducing issues of credibility and sustainability of fiscal policies.

The analysis of this chapter will allow us to answer questions such as:

- What is the role of fiscal policy in a monetary union?
- How independent can national fiscal policies be?
- Does a monetary union increase or reduce fiscal discipline? What rules, if any, should be used to restrict national fiscal policies? This will lead us into an analysis of the 'stability pact'.

1. Fiscal policies and the theory of optimum currency areas

In order to analyse what the theory of optimum currency areas has to say about the conduct of fiscal policies, it is useful to start from the example of an asymmetric demand shock as developed in Chapter 1. Suppose again that European consumers shift their demand in favour of German products at the expense of French products. We reproduce the figure of Chapter 1 here (Fig. 9.1). What are the fiscal policy implications of this disturbance?

Suppose first that France and Germany, being members of the same monetary union, have also centralized a substantial part of their national budgets to the central European authority. In particular, let us assume that the social security system is organized at the European level, and that income taxes are also levied by the European government.

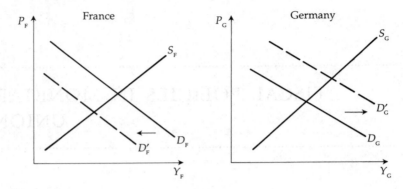

Fig. 9.1. Asymmetric shock in France and Germany

It can now be seen that the centralized budget will work as a shock absorber. In France output declines and unemployment tends to increase. This has a double effect on the European budget. The income taxes collected by the European government in France decline, whereas unemployment benefit payments by the European authorities increase. Exactly the opposite occurs in Germany. There, output increases and unemployment declines. As a result, tax revenues collected by the European government in Germany increase, while European government spending in Germany declines. Thus, the centralized European budget automatically redistributes income from Germany to France, thereby softening the social consequences of the demand shift for France.

Consider now what would happen if France and Germany form a monetary union without centralizing their government budgets. It can easily be demonstrated that the negative demand shock in France will lead to an increase in the French government budget deficit, because tax receipts decline while unemployment benefit payments by the French government increase. The French government will have to increase its borrowing. In Germany we have the reverse. The German government budget experiences increasing surpluses (or declining deficits). If capital markets work efficiently, the need for the French government to borrow can easily be accommodated by the increasing supply of savings coming from Germany.

In this case of decentralized budgets, France increases its external debt, which will have to be serviced in the future. This will reduce the degrees of freedom of future French fiscal policies. This contrasts with the case where the national budgets are centralized: in such a system, France will not have to face such external debt problems, as German residents automatically transfer income to France. They will not necessarily be repaid for their generosity.

It should also be stressed that the size of these budgetary effects very much depends on the degree of wage and price flexibility and/or labour mobility. If these are large, the automatic transfer from Germany (or, in the second case, the French budget deficit) will be low. If, for example, labour mobility between France and Germany is high, there will be little unemployment in France, so that unemployment benefit payments are correspondingly low.

The theory of optimum currency areas leads to the following implications for fiscal policies in monetary unions (see Kenen (1969)). First, it is desirable to centralize a significant part of the national budgets to the European level. A centralized budget allows countries (and regions) that are hit by negative shocks to enjoy automatic transfers, thereby reducing the social costs of a monetary union. This was also a major conclusion of the influential MacDougall Report, published in 1977. The drafters of that report argued that monetary union in Europe would have to be accompanied by a significant centralization of budgetary power in Europe (more precisely a centralization of the unemployment benefit systems). Failure to do so would impose great social strains and endanger monetary union. (In Box 12 we ask the question of how much centralization of national budgets is desirable.)

Secondly, if such a centralization of the national government budgets in a monetary union is not possible (as appears to be the case in the context of European monetary union), national fiscal policies should be used in a flexible way. That is, when countries are hit by negative shocks, they should be allowed to let the budget deficit increase through the built-in (or automatic) budgetary stabilizers (declining government revenues, increasing social outlays).

This requirement that fiscal policies respond flexibly to negative shocks also implies that a substantial autonomy should be reserved for these national fiscal policies. In the logic of the optimum currency area theory, countries lose an instrument of policy (the exchange rate) when they join the union. If there is no centralized budget which automatically redistributes income, countries have no instrument at their disposal to absorb the effects of these negative shocks.[1] The fiscal policy instrument is the only one left.

This theory about how fiscal policies should be conducted in a monetary union has been heavily criticized recently. This criticism is not directed at the first conclusion, i.e. that it is desirable to centralize a significant part of the national budgets in a monetary union. The criticism has been formulated against the second conclusion, which calls for flexibility and autonomy of national government budgets in monetary unions, when the degree of budgetary centralization is limited. To this criticism we now turn.

2. Sustainability of government budget deficits

The major problem with the previous analysis is the underlying assumption that governments can create budget deficits to absorb negative shocks without leading to problems of sustainability of these deficits. As many Western Eur-

---

[1] We assume here that wages are inflexible and that labour mobility is non-existent.

opean countries have experienced during the 1980s, however, government budget deficits can lead to such problems.

The sustainability problem can be formulated as follows. A budget deficit leads to an increase in government debt which will have to be serviced in the future. If the interest rate on the government debt exceeds the growth rate of the economy, a debt dynamic is set in motion which leads to an ever-increasing government debt relative to GDP. This becomes unsustainable, requiring corrective action.

This debt dynamics problem can be analysed more formally starting from the definition of the government budget constraint (see Box 9 for more explanation):

$$G - T + rB = dB/dt + dM/dt \tag{9.1}$$

where G is the level of government spending (excluding interest payments on the government debt), T is the tax revenue, r is the interest rate on the government debt, B, and M is the level of high-powered money (monetary base).

The left-hand side of equation (9.1) is the government budget deficit. It consists of the primary budget deficit $(G - T)$ and the interest payment on the government debt (rB). The right-hand side is the financing side. The budget deficit can be financed by issuing debt (dB/dt) or by issuing high-powered money dM/dt. (In the following we represent the changes per unit of time by putting a dot above a variable, thus $dB/dt = \dot{B}$ and $dM/dt = \dot{M}$.)

As is shown in Box 9, the government budget constraint (9.1) can be rewritten as

$$\dot{b} = (g - t) + (r - x)\, b - \dot{m} \tag{9.2}$$

where $g = G/Y$, $t = T/Y$, $x = \dot{Y}/Y$ (the growth rate of GDP), and $\dot{m} = \dot{M}/Y$.

Equation (9.2) can be interpreted as follows. When the interest rate on government debt exceeds the growth rate of GDP, the debt-to-GDP ratio will increase without bounds. The dynamics of debt accumulation can only be stopped if the primary budget deficit (as a percentage of GDP) turns into a surplus $((g - t)$ then becomes negative). Alternatively, the debt accumulation can be stopped by a sufficiently large revenue from money creation. The latter is also called seigniorage. It is clear, however, that the systematic use of this source of finance will lead to inflation.

The nature of the government budget constraint can also be made clear as follows: one can ask the question under what condition the debt-to-GDP ratio will stabilize at a constant value. Equation (9.2) gives the answer. Set $\dot{b} = 0$. This yields

$$(r - x)\, b = (t - g) + \dot{m}. \tag{9.3}$$

Thus, if the interest rate exceeds the growth rate of the economy, it is necessary that either the primary budget shows a sufficiently high surplus $(t > g)$ or that money creation is sufficiently high in order to stabilize the debt–GDP ratio.

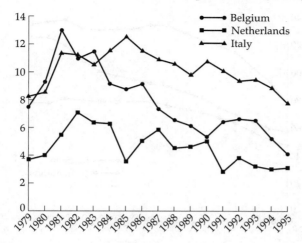

Fig. 9.2. General government budget deficit (per cent of GDP)
Source: EC Commission, *European Economy*, 1995.

The latter option has been chosen by many Latin American countries during the 1980s, and more recently by some Eastern European countries. It has also led to hyperinflation in these countries.

The important message here is that, if a country has accumulated sizeable deficits in the past, it will now have to run correspondingly large primary budget surpluses in order to prevent the debt-to-GDP ratio from increasing automatically. This means that the country will have to reduce spending and/or increase taxes.

It may be interesting here to present a few case-studies of the experience of European countries during 1980–95. We select Belgium, the Netherlands, and Italy. We present data on government budget deficits of these countries during 1980–95 (see Fig. 9.2).

It can be seen that these countries allowed their budget deficits to increase significantly during the early part of the 1980s. This increase in government budget deficits came mainly as a response to the negative consequences of major recessions in these countries. The rest of the period was characterized by strenuous attempts to contain the explosive debt situation that followed from these earlier policies. We show the evolution of the debt-to-GDP ratios in these countries during 1980–95 in Fig. 9.3. At the end of the 1980s after many years of budgetary restrictions, Belgium and the Netherlands had succeeded in stabilizing the debt-to-GDP ratio. They were able to do so by running substantial surpluses in the primary budget. These are shown in Fig. 9.4. The level at which Belgium achieved this stabilization of the debt-to-GDP ratio, however, was very high. As a result, when the next recession hit the country in 1992–3 it was quite difficult to use fiscal policies in a countercyclical way. On

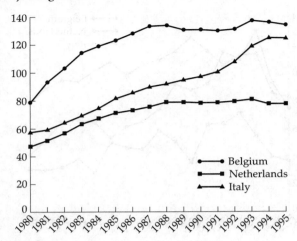

Fig. 9.3. Gross public debt (per cent of GDP)

Source: EC Commission, *European Economy*, 1995.

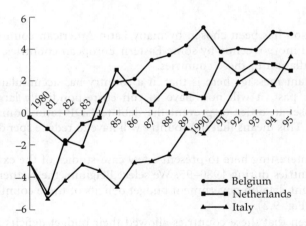

Fig. 9.4. Government budget surplus, excluding interest payments (per cent of GDP)

Source: EC Commission, *European Economy*, 1995.

the contrary, Belgium was forced to clamp down on government spending and to increase taxes, thereby exacerbating the recession, without preventing the debt–GDP ratio from increasing again during 1992–93.

Italy did not succeed in stabilizing its debt–GDP ratio during the 1980s because it failed to create the required primary budget surpluses. Only from 1992 on did it manage to produce primary budget surpluses that led to a stabilization of the debt–GDP ratio in 1994–5. Note that this prevented the Italian government from using budgetary policies in a countercyclical way.

These examples vividly demonstrate the limits to the use of fiscal policies to offset negative economic shocks. Such policies cannot be maintained for very long. The experience of these countries shows that large government budget deficits quickly lead to an unsustainable debt dynamics from which countries find it difficult to extricate themselves. The 'success' of Belgium in stabilizing the debt–GDP ratio by running primary budget surpluses during the second half of the 1980s came about after many years of spending cuts and tax increases.

The previous discussion also makes clear that fiscal policies are not the flexible instrument that the optimum currency theory has made us believe. The systematic use of this instrument quickly leads to problems of sustainability, which forces countries to run budget surpluses for a number of years. Put differently, when used once, it will not be possible to use these fiscal policies again until many years later.

This analysis of the sustainability of fiscal policies has led to a completely different view of the desirable fiscal policies of member states in a monetary union. This view found its reflection in the Maastricht Treaty, which defines budgetary rules countries have to satisfy in order to enter EMU (the 3% deficit and the 60% debt norms).[2] It also found expression in the so-called 'stability pact' that the EU heads of state, at the insistence of Germany, agreed would have to be implemented after the start of EMU. This stability pact is quite important as it is likely to guide national fiscal policies in the future EMU. Its main principles are the following. First, countries will have to aim at achieving balanced budgets. Second, countries with a budget deficit exceeding 3% of GDP will be subject to fines. These fines can reach up to 0.5% of GDP. Third, these fines will not be applied if the countries in question experience exceptional circumstances, i.e. a natural disaster or a decline of their GDP of more than 2% during one year. In cases where the drop in GDP is between 0.75 and 2% the application of the fine will be subject to the approval of the EU finance ministers. Countries that experience a drop in their GDP of less than 0.75% have agreed not to invoke exceptional circumstances. The presumption is that in that case they will have to pay a fine, although even in that case the imposition of the fine will require a decision of the Council.[3]

The Maastricht Treaty and the stability pact take the view that fiscal policies in a monetary union should be subjected to rules. Let us evaluate the arguments for rules on government budget deficits in a monetary union.

---

[2] The drafters of the Maastricht Treaty were very much influenced by the Delors Report, which was the first to express the need for strict rules on budgetary policies.

[3] The original proposals of the German minister of finance implied the automatic application of fines. This, however, would have been in contradiction to the Maastricht Treaty, which stipulates that fines can only be imposed with a majority of two-thirds of the weighted votes in the Council. This implies that the implementation of the fines foreseen in the stability pact will in any case need a two-thirds majority independent of the question whether or not the excessive deficit came about because of a drop in GDP, and whatever the size of that drop.

3. The argument for rules on government budget deficits

The basic insight of this view is that a country that finds itself on an unsustainable path of increasing government debt creates negative externalities for the rest of the monetary union. A country that allows its debt–GDP ratio to increase continuously will have increasing recourse to the capital markets of the union, thereby driving the union interest rate upwards. This increase in the union interest rate in turn increases the burden of the government debts of the other countries. If the governments of these countries choose to stabilize their debt–GDP ratios, they will be forced to follow more restrictive fiscal policies. Thus, the unsustainable increase in the debt of one country forces the other countries to follow more deflationary policies. It will therefore be in the interest of these other countries that a control mechanism should exist restricting the size of budget deficits in the member countries.

There is a second externality that may appear here. The upward movement of the union interest rate, following the unsustainable fiscal policies of one member country, is likely to put pressure on the ECB. Countries that are hurt by the higher union interest rate may pressure the ECB to relax its monetary policy stance. Thus, unsustainable fiscal policies will interfere with the conduct of the European monetary policy. Again it may be in the interest of the members of the union to prevent such a negative externality to occur by imposing limits on the size of government budget deficits.

These arguments based on the spillover effects of fiscal policies appear reasonable. They have, however, been subjected to serious criticism.[4] The criticism has been twofold. One is theoretical and concerns the role of capital markets. The second has to do with the enforceability of such rules.

(*a*) *The efficiency of private capital markets.* Implicit in the externality argument, there is an assumption that capital markets do not work properly. Let us now suppose that capital markets work efficiently, and ask the question what will happen when one country, say Italy, is on an unsustainable debt path. Does it mean that the union interest rate will increase, i.e. that the interest rate to be paid by German, Dutch, or French borrowers will equally increase? The answer is negative. If capital markets in the monetary union work efficiently, it will be recognized that the debt problem is an Italian problem. The market will attach a risk premium to Italian government debt. The German government, however, will not be affected by this. It will be able to borrow at a lower interest rate, because the lenders recognize that the risk inherent in German government bonds is lower than the risk involved in buying Italian government debt instruments. Thus, if the capital markets work efficiently, there will be no externality. Other governments in the union will not suffer from the existence of a high Italian government debt. In addition, it does not make sense to talk

---

[4] See e.g. Buiter and Kletzer (1990), van der Ploeg (1990), von Hagen (1990), and Wyplosz (1991).

about *the* union interest rate. If capital markets are efficient there will be different interest rates in the union, reflecting different risk premia on the government debt of the union members.

This is certainly a powerful argument. There is, however, a possibility that lenders will find it difficult to attach the correct risk premium to the Italian government debt. Let us analyse under what conditions this may happen. Suppose the lenders believe that in case of a serious debt crisis, i.e. an inability of the Italian government to service its debt, the other countries will step in to 'bail out' the Italian government. These countries may have an interest in doing so, because an Italian debt crisis may spill over to the rest of the financial system. For example, financial institutions in other countries may hold Italian government paper. A default by Italy may lead to defaults of these financial institutions, and may create a general debt crisis. To avoid this, the governments of other countries may decide to step in and buy up the Italian government paper. The realization that such an implicit bailout guarantee exists will lower the risk premium on Italian government paper. Thus, the capital market fails to attach the correct price to risky Italian debt instruments.

One way to solve this problem consists in a solemn declaration of the members of the union that they will never bail out other members countries' governments. Such a 'no-bail-out' clause was in fact introduced in the Maastricht Treaty. Although such a clause may help to resolve the problem, it is unlikely completely to eliminate it because it may not be fully credible. When an Italian debt crisis occurs, it will be in the interest of the other member countries to bail out the Italian government for the reasons given earlier. Thus, even if they have previously declared their intention not to intervene, they may very well decide to make an exception this time. The realization that this may happen in the future will hamper private lenders in correctly pricing the risk of Italian government debt.

(*b*) *The enforcement of fiscal policy rules.* A second problem with rules on the size of government budget deficits and debts has to do with the enforceability of these rules. Experience with such rules is that it is very difficult to enforce them. A recent example of such difficulties is the Gramm–Rudman legislation in the USA. In 1986 the US Congress approved a bill that set out explicit targets for the US Federal budget deficit. If these targets were not met, spending would automatically be cut across the board by a given percentage so as to meet the target. It can now be said that this approach with rules was not very successful. The US executive branch found all kinds of ways of circumventing this legislation. For example, some spending items were put 'off the budget'.

There is also evidence collected by von Hagen (1991) for the US states pointing in the same direction. Von Hagen found that those states that had constitutional limits on their budget deficits or on the level of their debt had frequent recourse to the technique of 'off-budgeting'. As a result, he found that the existence of constitutional rules had very little impact on the size of the states' budget deficits.

Similarly, as European countries approach 1999, when they will be forced to comply with the Maastricht budgetary rules, they are relying more and more on 'creative budgetary accounting' to camouflage their true debts and deficits. It is to be expected that this will continue in the future.

To sum up, one can say that some rules on the conduct of fiscal policies by the EMU members will be necessary. Given the interdependence in the risk of bonds issued by different governments, financial markets may find it difficult to price these risks correctly. As a result, some form of mutual control will be necessary. The question arises, however, whether the stability pact may not have gone too far in stressing rigid rules on the conduct of fiscal policies

Before coming to a final verdict, it is important to analyse two issues. First, we study the extent to which monetary unions impose additional discipline on national budgetary authorities. If so, the need for additional rules is weakened. Second, we analyse the risk of defaults and bailouts in monetary unions. This risk is often considered to provide the major rationale for additional rules on fiscal policies in a monetary union.

4. Fiscal discipline in monetary unions

An important issue relating to the need for rules on fiscal policies has to do with the way a monetary union affects the fiscal discipline of national governments in such a union. Proponents of rules have generally argued that a monetary union is likely to reduce the fiscal discipline of national governments. Adversaries of rules have argued the opposite.

In order to see clearly the different, and opposing, arguments it is useful to ask the question of how a monetary union may change the incentives of fiscal policy-makers, and, in so doing, may affect budgetary discipline.

The issue of whether a monetary union increases or reduces the degree of fiscal discipline of countries joining the union has been hotly debated in the literature.[5] Broadly speaking there are two factors that are important here and that can lead to a change in the incentives of countries in regard to the size of their budget deficits when they join the union. The first one leads to less discipline, the second to more discipline.

The phenomenon that leads to incentives for larger budget deficits can be phrased as follows. When a sovereign country issues debt denominated in the domestic currency, the interest rate it will have to pay reflects a risk premium consisting of two components, the risk of default and the risk that the country will devalue its currency in the future. The latter can be completely eliminated by issuing debt in foreign currency. The risk premium then reflects the pure default risk (sometimes also called credit risk). This also happens when a country joins a monetary union: the governments of the member states have

---

[5] For a survey see Wyplosz (1991).

Table 9.1. Budget deficit of member states of unions and of EC member countries (as a per cent of revenues)

	Weighted mean	Unweighted mean
USA (1985)	+10.9	+4.6
Australia (1986–7)	−10.1	−9.1
West Germany (1987)	−6.4	−8.2
Canada (1982)	−0.4	−1.4
Switzerland (1986)	−1.3	−0.7
European Community (1988)	−10.1	−11.4

Note: A positive sign is a surplus, a negative sign a deficit. The weighted mean is obtained by weighting the deficits with the share of the member state in the union's GDP.
Source: Lamfalussy (1989: 102–24)

to issue debt in what is equivalent to a 'foreign' currency. As a result, the risk premium (if any) will reflect the probability that the issuing government may not fully service its debt in the future.

As was indicated earlier, however, the lenders may have difficulties in correctly estimating this default risk because of an implicit bailout guarantee that other members of the union extend. The possibility of a bailout then gives an incentive to member states to issue unsustainable amounts of debt. A no-bailout provision may not solve this problem because it is not likely to be credible. Even if the European authorities were solemnly to declare never to bail out member states, it is unsure whether they would stick to this rule if a member country faced the prospect of being unable to service its debt. Thus, the monetary union leads to excessive budget deficits of the member states.

There is a second factor, however, which tends to reduce the incentive of member states of a monetary union to run excessive deficits. Countries who join the union reduce their ability to finance budget deficits by money creation. As a result, the governments of member states of a monetary union face a 'harder' budget constraint than sovereign nations. The latter are confronted with 'softer' budget constraints and therefore have a stronger incentive to run deficits.[6]

Which one of the two effects—the moral hazard or the no-monetization one—prevails is essentially an empirical question in that it depends on institutional features and on the incentives governments face. It is, therefore, useful to analyse the experience of member states in existing monetary unions and to compare this with the experience of sovereign nations. This is done in Table 9.1, where the average budget deficits of member states of existing monetary unions are presented together with the average of the budget deficits of the EC countries.

The most striking feature of Table 9.1 is the fact that the *average* budgetary deficits of the member states in monetary unions tend to be lower than the

---

[6] For a formalization of the incentives of governments in multi-party systems, see Alesina and Tabellini (1987) and Persson and Svensson (1989). A classic analysis is Buchanan and Tullock (1962).

average deficit of independent countries in the EC.[7] This suggests that on average the existence of a union adds constraints to the size of the budget deficits of the member states. Thus, it appears that the no-monetization constraint is a powerful disincentive against running large budget deficits. It therefore seems that governments of member states of a monetary union face a 'harder' budget constraint than sovereign nations. The latter are confronted with 'softer' budget constraints and therefore have a stronger incentive to run deficits.

It should also be noted that, except in the case of Germany, in none of the monetary unions analysed in Table 9.1 does the federal authority impose restrictions on the budget deficits of the member states. In the Federal Republic the federal government can limit the borrowing requirements of the *Länder* with the consent of their representatives in the Bundesrat. This provision, however, has been invoked only twice (in 1971 and 1973).[8]

Whereas federally imposed limits are rare, one frequently finds *self-imposed* constitutional limits on state and *Länder* budget deficits. In fact, in the USA the majority of the states have introduced such constitutional limitations. One possible interpretation for the frequency of self-imposed limits is that they may help to improve the reputation and the creditworthiness of the states in the capital markets, and therefore reduce the cost of borrowing.[9]

The evidence of Table 9.1 is of course not conclusive. More research should be carried out to find out whether the difference observed between the average deficits of member states of monetary unions and the average deficits of sovereign nations can be generalized to different time-periods and a larger sample of countries. At the very least, the results of Table 9.1 suggest that the idea that in monetary unions member states have a strong incentive to create excessive levels of government debt is not corroborated by the facts of the 1980s.

Additional indirect evidence for the hypothesis that the size of deficits depends on the 'softness' of the budget constraint is provided by Moesen and Van Rompuy (1990). They classify industrial countries according to the degree of centralization of total government spending. The hypothesis tested is that more centralized governments face a softer budget constraint than decentralized governments. This is so because in centralized countries a larger part of total government spending can potentially be financed by the issue of money. In decentralized countries, however, a larger part of government spending occurs through lower-level authorities who lack the access to monetary financing. Moesen and Van Rompuy find evidence for this hypothesis: during the period 1973–86 the total government deficits in centralized countries increased

[7] This is also confirmed by Van Rompuy *et al.* (1991).

[8] Van Rompuy *et al.* (1991: 19).

[9] As discussed earlier, von Hagen (1990) has argued that these constitutional limits on the budget deficits have been relatively ineffective. This would then suggest that, if perceived as ineffective in the capital markets, these constitutional restrictions should have a low effect on the risk premium.

faster than in decentralized ones. Put differently, during a period when government budgets were negatively affected by supply shocks and economic recession, the countries with more decentralized governments faced a stronger pressure to reduce spending and/or increase taxes than countries with more centralization of government functions.[10]

5. Risks of default and bailout in a monetary union

The discussion of the previous section allows us to shed some new light on the issue of whether a debt default becomes more likely in a monetary union, so that the risk of having to organize a costly bailout also increases.

From our previous discussion one might conclude that since the member states of the EMU will not behave in a less disciplined way than they do today outside a monetary union, the risk of default should not increase in a monetary union.

There are other considerations, however, that one should add here to arrive at a correct appreciation of the risk of default in a future monetary union. One has been stressed recently by McKinnon (1996), the other by Eichengreen and von Hagen (1995).

Sovereign nations can default on their debt in two ways. One is an outright default (e.g. stopping payment of interest on the outstanding debt). The other is an implicit default by creating surprise inflation and devaluation, which reduces the real value of the debt. Sovereign nations who control their own national banks can always resort to surprise inflation and devaluation to reduce the burden of their debt. They often do this to avoid outright default.

As soon as a country joins a monetary union it loses control over the central bank, and therefore cannot create surprise inflation to reduce the burden of its debt any more. As a result, pressure on the government to organize an outright default may actually increase in a monetary union. McKinnon (1996) argues that this will happen in EMU. The level of the debt of certain EU countries is so high that in the absence of implicit defaults by inflation and devaluation, the probability that outright default will occur increases.

Differences in interest rates allow us to obtain an idea about the nature of the risks involved. In Table 9.2 we present interest rate spreads for long-term bonds between a number of EU countries and Germany. The first column presents the total spread. For example, the spread between a ten-year lira bond issued by the Italian government and a ten-year DM bond issued by the German government amounted to 444 basis points (4.44%) in March 1996. This differential can be interpreted as a premium the Italian government had

[10] Roubini and Sachs (1989) have identified other institutional features that have led to different trends in budget deficits, i.e. the degree of political cohesion of the governments in power.

Table 9.2. Interest differentials with Germany (ten-year bonds), March 1996

	in national currencies (1)	in common currency (DM) (2)	(3) = (1) − (2)
Guilder	4	5	−1
French franc	37	21	16
Belgian franc	50	27	23
Danish krone	129	11	118
Pound sterling	160	9	151
Peseta	347	30	317
Lira	444	81	363

Source: JP Morgan, March 1996.

to pay for the risk that the lira may be devalued (implicit default) plus the risk of an outright default. One can disentangle this risk premium as follows. We compute the interest differential on ten-year bonds issued by the Italian and German governments *in the same currency* (the DM). This is given in the second column. This differential can be interpreted as the premium for the risk of outright default (credit risk). The third column, then, is the difference between the first two columns and expresses the devaluation risk (risk of implicit default).

We observe the following interesting phenomenon. The premia for the risk of outright default (column (2)) is relatively small for all currencies, never exceeding 100 basis points (1%). The premia for the risk of devaluations (column (3)) is much higher for many currencies. Thus the market believes that governments are more likely to follow policies leading to implicit rather than outright default. When the currencies involved join the monetary union, the devaluation risk will suddenly drop to zero. For example, the Italian bonds will be converted into the common currency, the euro, so that the Italian authorities can no longer depreciate the currency in which they issue their bonds. Will this increase the premia for the risk of outright default? According to McKinnon, it will. The evidence provided by the American states seems to support this. For we observe that the interest differentials on the bonds issued by American states, and expressed in the same currency, the dollar, are significantly higher than the spreads observed in column (2) of Table 9.2 This suggests that part of the devaluation risk of currencies like the lira and the peseta will become an outright default risk when countries join the EMU.

It is very unlikely, however, that this substitution between devaluation risk and credit risk will be complete. That is, it is very unlikely that the interest differential between, say, Italian and German bonds in the future EMU (which will be expressed in the same currency, the euro) will be as large as today's differential between lira and DM bonds (column (1) in Table 9.2). Thus, Italy is likely to profit from lower risk premia once it is in the EMU. This by itself tends to reduce the risk of default.

This conclusion is buttressed by the research of Eichengreen and von Hagen

(1995). These authors argue, and provide evidence, that member countries of a monetary union who maintain control over a large domestic tax base face a low default risk compared to members of a monetary union with few fiscal responsibilities. Since the European monetary union will consist of countries maintaining large domestic taxing powers, the risk of default is likely to be small, compared to the risk faced by, say, the American states or the Canadian provinces which, compared to the EU countries, have limited taxing powers. Eichengreen and von Hagen conclude from this analysis that the need to impose tight rules on the government budgets of member countries in the EMU has been overemphasized.

Default risks will never be zero, however, so that the risk of having to organize a costly bailout will also never be zero. The question then is whether the risk of bailouts is larger when countries are in the EMU than when they stay outside. Let us suppose that Italy defaults on its debt. Is this more likely to lead to a bailout operation if Italy is in the EMU than if it is left outside the EMU?

The standard argument for an affirmative answer runs as follows. In a monetary union, financial integration increases. As a result, bonds issued by the different national authorities will be more widely distributed across different member countries. Therefore, when one government defaults on its debt this will affect more individuals and financial institutions outside the defaulting country than if the country had not been in the union. The result is that pressure exerted on the other governments to bail out the defaulting government will be stronger when that government is in the union than when it stays outside.

Although it cannot be denied that the strong financial integration of a monetary union provides the potential for a lot of pressure to bail out defaulting governments, this is not the only relevant consideration. There is also the exchange rate issue. When a country, say Italy, is not allowed in the union one can expect that a default will also put a lot of pressure on the other EU members to bail out the Italian government. This pressure comes from the fact that if Italy is outside the monetary union when it defaults, the lira is likely to collapse in the foreign exchange market, causing industrialists in the rest of the EU to put a lot of pressure on their governments to support the lira. This exchange rate effect is absent if Italy defaults when it is a member of the monetary union. We conclude that keeping Italy outside the union does not necessarily reduce the risk that EU members will have to engage in a future bailout operation. The two effects, the financial integration effect and the exchange rate effect, operate in opposite directions. We simply do not know a priori whether in a monetary union the pressure to bail out a defaulting EU country will be stronger than when this same EU country is left outside the EMU.

6. The stability pact: an evaluation

National fiscal policies in the future EMU must find a balance between two conflicting concerns. The first one has to do with flexibility and is stressed in the theory of optimum currency areas: in the absence of the exchange rate instrument and a centralized European budget, national government budgets are the only available instruments for nation-states to confront asymmetric shocks. Thus, in the future EMU national budgets must continue to play some role of automatic stabilizers when the country is hit by a recession.

A second concern relates to the spillover effects of unsustainable national debts and deficits, which were described in the previous sections. Unsustainable debts and deficits in particular countries may harm other member countries and may exert undue pressure on the ECB.

How does the stability pact strike a balance between these two concerns? It is clear that the stability pact has been guided more by the fear of unsustainable debts and deficits than by the need for flexibility. As a result, it is fair to say that the stability pact is quite unbalanced in stressing the need for strict rules at the expense of flexibility. This creates a risk that the capacity of national budgets to function as automatic stabilizers during recessions will be hampered, thereby intensifying recessions. We illustrate the nature of this risk by analysing what happened during the recession of the early 1990s in the EU countries (see Fig. 9.5). We show the increase of budget deficits in the EU countries during the recession of the early 1990s.[11] At the same time Fig. 9.5 presents the largest yearly decline in GDP reached during that recession. We observe that six countries (Finland, France, Sweden, Spain, Portugal, and the UK) saw their budget deficit increase by more than 3% during the recession. Some of them (Finland, Sweden, and the UK) would have been able to invoke exceptional circumstances (a decline of GDP by more than 2% for a year) and would have been spared a penalty. The other three countries (France, Portugal, and Spain) would not have been able to invoke these exceptional circumstances, because GDP decline never exceeded 2% a year. These countries therefore would have faced the possibility of being subjected to fines. This could have influenced their behaviour and led them to apply more restrictive fiscal policies during the recession.

From the evidence of Fig. 9.5 it follows that increases in the budget deficit of more than 3% of GDP are not uncommon during recessions. This also happens in countries that experience recessions involving a decline in their GDP of less than 2% (the benchmark for avoiding fines in the stability pact). Since the stability pact sets the maximum limit of budget deficits to 3% of GDP, this implies that countries will have to run government budget surpluses

---

[11] Note that the timing of the recession is not the same in all countries. In the UK the steepest decline in GDP occurred in 1990, in Sweden and Finland in 1991, and in the other EU countries in 1993.

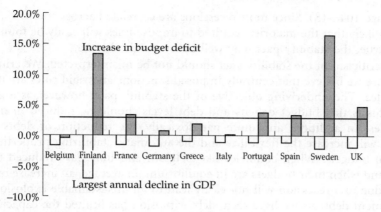

Fig. 9.5. Increase in budget deficits during 1991–1993

Sources: IMF, *International Financial Statistics*; EC Commission, *European Economy*.

on average in order to have sufficient flexibility during recessions, i.e. in order to avoid hitting the ceiling of 3% of GDP.

The lack of budgetary flexibility to face recessions will create tensions between national governments and European institutions. This tension will exist at two levels. First, as countries will be hindered in their desire to use the automatic stabilizers in their budgets during recessions, they will increase their pressure on the ECB to relax monetary policies. Thus, paradoxically, the stability pact whose aim it was to protect the ECB from political pressure may in fact have increased the risk of such pressure. Second, when countries are hit by economic hardship, EU institutions will be perceived as preventing the alleviation of the hardship of those hit by the recession. Worse, they will be seen handing out fines and penalties when countries struggle with economic problems. This will certainly not promote enthusiasm for European integration. On the contrary, it is likely to intensify Euro-scepticism.

We conclude that the stability pact has gone too far in imposing rules on national government budgets. The lack of flexibility of future national budgetary policies in the EMU will create risks that are larger than the risks of default and bailouts stressed by the proponents of rules. As we argued in the previous sections, there is very little evidence that a monetary union increases fiscal indiscipline and the risks of default and bailouts compared to a situation without a monetary union.

One can of course argue that the stability pact will not function, and therefore should not worry us unduly. There is indeed a serious possibility that the implementation of the stability pact will be very difficult. We mentioned earlier that in countries where rules on budgetary policies are imposed, all kinds of creative accounting emerge. In addition, any imposition of fines will have to be decided by a majority of two-thirds in the Council (Maastricht

Treaty art. 104c–13). Since most recessions are correlated across countries, it is very unlikely that the majority required to impose fines will easily be found. If this is true, the stability pact may well become a dead letter.

Our criticism of the stability pact should not be misinterpreted. We criticize it because we believe that centrally imposed sanctions and rigid rules are not a good idea. The underlying objective of the stability pact, however, is a good one. This is that budget deficits and debt levels should be reduced in many countries. In addition, and quite paradoxically, the reduction of debts and deficits will increase the flexibility and the automatic stabilizing properties of national budgets. When countries like Belgium and Italy have reduced their debts, and when their budgets are in equilibrium on average, an increase in the deficit due to a recession will not easily lead to an unsustainable explosion of government debt. As we have seen, debt explosion has limited the capacity of these nations to use automatic stabilizers during the recent recessions. Thus, restoring fiscal rectitude creates the conditions for using the budget as a tool for absorbing asymmetric shocks and for stabilizing the economy. In that sense fiscal orthodoxy is badly needed to increase the flexibility of a monetary union.

Box 12. How much centralization of government budgets in a monetary union?

The theory of optimum currency areas stresses the desirability of a significant centralization of the national budgets to accommodate for asymmetric shocks in the different regions (countries). What is the limit to such centralization? In order to answer this question it is important to realize that budgetary transfers should be used to cope only with *temporary* shocks, or, when shocks are permanent, these transfers should be used only temporarily. A country or a region that faces a permanent shock (e.g. a permanent decline in the demand for its output) should adjust by wage and price changes or by moving factors of production. The budgetary transfers can be used only temporarily to alleviate these adjustment problems.

The experience with regional budgetary transfers within nations (Italy and Belgium, for example), however, is that it is very difficult to use these transfers in a temporary way. Quite often, when a region experiences a negative shock (the Mezzogiorno in Italy, Wallonia in Belgium), the transfers through the centralized social security system tend to acquire a permanent character. The reason is that these social security transfers reduce the need to adjust. They tend to keep real wages in the depressed regions too high and they reduce the incentive of the population of the region to move out to more prosperous regions. As a result, these transfers tend to become self-perpetuating. This is illustrated by the fact that the Mezzogiorno has been a recipient of transfers (from the rest of Italy) representing 20–30% of the Mezzogiorno's regional output during most of the last twenty-five years.[12] Although less important in size, similar regional transfers exist in other countries (e.g. Belgium).

These large and permanent regional transfers then create new political pro-

---

[12] See Micossi and Tullio (1991).

blems, when the inhabitants of the prosperous regions increasingly oppose paying for them. In some countries these political problems can even lead to calling into question the unity of the nation. When the sense of national identity is weak, this can effectively lead to a break-up of the country.

The experience with regional transfers within European nations is important to bear in mind when considering the limits that should be imposed on the centralization of national budgets (including social security) in Europe. Surely, a centralization of the social security system at the European level would almost certainly create Mezzogiorno problems involving whole countries. This would lead to quasi-permanent transfers from one group of countries to another. The sense of national identification being much less developed at the European level than at the country level, this would certainly lead to great political problems. These would in turn endanger the unity of the European Union. Thus, although Europe needs some further centralization of the national budgets (including the social security systems) to have a workable monetary union, the degree of centralization should stop far short from the level achieved within the present-day European nations. Recently some schemes of limited centralization of the social security systems (in particular, the unemployment benefit systems) have been proposed and worked out by a number of economists (see Italianer and Vanheukelen (1992), EC Commission (1993), Hammond and von Hagen (1993), and Mélitz and Vori (1993)). These proposals suggest that a limited centralization can be quite effective in taking care of (temporary) asymmetric shocks.

7. Conclusion

According to the theory of optimum currency areas, a monetary union in Europe should go together with some centralization of the national budgets. Such a centralization of the budgetary process allows for *automatic* transfers to regions and countries hit by negative shocks.

It is quite unlikely that a significant centralization of the national budgets will be achieved in the near future in the Community. For some observers, in particular those who published the MacDougall Report, this means that it would be advisable to postpone further steps towards monetary unification.

If monetary unification in Europe is realized, this will be done despite the fact that no significant central European budget exists. This poses the question of how national fiscal policies should be conducted in a monetary union.

In this chapter we discussed two views about this problem. The first one is based on the theory of optimum currency areas and suggests that national fiscal authorities should maintain a sufficient amount of flexibility and autonomy. The second found its reflection in the Maastricht Treaty and the 'stability pact'. According to this view, the conduct of fiscal policies in the future monetary union will have to be disciplined by explicit rules on the size of the national budget deficits.

We have evaluated these two views. The optimum currency area view is probably overly optimistic about the possibility of national budgetary authorities using budget deficits as instruments to absorb negative shocks. Although there are situations in which countries will need the freedom to allow the budget to accommodate for these negative shocks, the sustainability of these policies limits their effectiveness.

We also argued, however, that the case for strict rules on the size of national government budget deficits is weak. There is no evidence that these rules are enforceable. In addition, the fact that national governments in a future monetary union will not have the same access to monetary financing as most of them have today 'hardens' the budget constraint and reduces the incentives to run large budget deficits. The fear that national authorities will be less disciplined in a monetary union than in other monetary regimes does not seem to be well founded.

From the discussion of this chapter it will be clear that there are many unresolved issues concerning the design and the use of fiscal policies in a monetary union. There are two issues worth mentioning here. First, the future monetary union in Europe, if it comes about, will most likely have to operate without a centralized European budget of a significant size. As a result, there will be no automatic mechanism redistributing income between the regions. Assuming that labour mobility between countries will remain small, the adjustment mechanism needed to deal with asymmetric shocks will take the form of relative price changes. It is unclear today whether this mechanism will be flexible enough to deal with such disturbances. We cannot exclude the possibility that some countries will find it difficult to adjust.

Secondly, many issues concerning the interaction between the European central bank and the national fiscal authorities will have to be resolved. It is one thing to legislate that the national authorities will have no access to monetary financing. It is quite another to enforce such a rule. When large countries are affected by negative disturbances, the authorities of these countries are likely to pressure the European central bank to follow accommodative policies. Thus, the absence of a reliable mechanism able to deal with asymmetric shocks is likely to put pressure on the European monetary authorities. This may impose an inflationary bias on the future monetary union in Europe. In order to withstand this pressure, the ECB will need institutional guarantees sufficient for maintaining its political independence.

The decision to go ahead with monetary union has clearly been inspired by the political objective of European unification. In this drive towards political union, the economic arguments both in favour of and against monetary union have often played a secondary role. They are important, however, because a failure to take them into account may lead to disillusions and political tensions. Only the future will tell for sure whether the decision to form a monetary union in Europe will be beneficial for all countries concerned.

REFERENCES

Aizenman, J., and Frenkel, J. (1985) 'Optimal Wage Indexation, Foreign Exchange Intervention, and Monetary Policy', *American Economic Review*, 75: 402–23.

Akerlof, G., Dickens, W., and Perry, G. (1996) 'The Macroeconomics of Low Inflation', Brookings Papers on Economic Activity, no. 1, 1–76.

Alesina, A. (1989) 'Politics and Business Cycles in Industrial Democracies', *Economic Policy*, 8: 55–98.

—— and Grilli, V. (1993) 'On the Feasibility of a One- or Multi-Speed European Monetary Union', NBER Working Paper, no. 4350.

—— and Summers, L. (1993) 'Central Bank Independence and Macroeconomic Performance: Some Comparative Evidence', *Journal of Money, Credit and Banking*, 25.

—— and Tabellini, G. (1987) 'A Positive Theory of Fiscal Deficits and Government Debt in a Democracy', NBER Working Paper, no. 2308.

'All Saints' Day Manifesto', *The Economist*, 1 Nov. 1975.

Artis, M. J., and Taylor, M. P. (1988) 'Exchange Rates, Interest Rates, Capital Controls and the European Monetary System: Assessing the Track Record', in F. Giavazzi *et al.* (1988).

—— and Zhang, W. (1995) 'International Business Cycles and the ERM: Is There a European Business Cycle?' CEPR Discussion Paper, no. 1191.

Backus, D., and Driffill, J. (1985) 'Inflation and Reputation', *American Economic Review*, 75: 530–8.

Bade, R., and Parkin, M. (1978) 'Central Bank Laws and Monetary Policies. A Preliminary Investigation: The Australian Monetary System in the 1970s', Clayton: Monash University.

Balassa, B. (1961) *The Theory of Economic Integration*, London: Allen & Unwin.

Baldwin, R. (1989) 'On the Growth Effects of 1992', *Economic Policy*, 11.

Barro, R. (1972) 'Inflationary Finance and the Welfare Cost of Inflation', *Journal of Political Economy*, 80: 978–1001.

—— and Gordon, D. (1983) 'Rules, Discretion and Reputation in a Model of Monetary Policy', *Journal of Monetary Economics*, 12: 101–21.

Bayoumi, T., and Eichengreen, B. (1993) 'Shocking Aspects of European Monetary Integration', in F. Torres and F. Giavazzi (eds.), *Adjustment and Growth in the European Monetary Union*, London: CEPR, and Cambridge: CUP.

—— —— (1996) 'Operationalizing the Theory of Optimum Currency Areas', CEPR Discussion Paper, no. 1484.

—— and Masson, P. (1994) 'Fiscal Flows in the United States and Canada: Lessons for Monetary Union in Europe', CEPR Discussion Paper, no. 1057.

—— and Prassad, E. (1995) 'Currency Unions, Economic Fluctuations and Adjustment: Some Empirical Evidence', CEPR Discussion Paper, no. 1172.

Begg, D., and Wyplosz, C. (1987) 'Why the EMS? Dynamic Games and the Equilibrium

Policy Regime', in R. Bryant and R. Portes (eds.), *Global Macroeconomics: Policy Conflict and Cooperation*, New York: St. Martin's Press.

—— Chappori, P., Giavazzi, F., Mayer, C., Neven, D., Spaventa, L., *et al.* (1991) *Monitoring European Integration: The Making of the Monetary Union*, London: CEPR.

—— Giavazzi, F., von Hagen, J., and Wyplosz, C. (1997) *EMU. Getting the End-game Right*, London: CEPR.

Bertola, G., and Svensson, L. (1993) 'Stochastic Devaluation Risk and the Empirical Fit of Target-Zone Models', *Review of Economic Studies*, 60: 689–712.

Bini-Smaghi, L., and Vori, S. (1993) 'Rating the EC as an Optimal Currency Area: Is it Worse than the US?', Banca d'Italia Discussion Paper, no. 187.

—— Padoa-Schioppa, T., and Papadia, F. (1993) 'The Policy History of the Maastricht Treaty: The Transition to the Final Stage of EMU', Banca d'Italia.

Blanchard, O., and Muet, P.-A. (1993) 'Competitiveness through Disinflation: An Assessment of the French Macroeconomic Strategy', *Economic Policy*, 16: 11–56.

Boyd, C., Gielens, G., and Gros, D. (1990) 'Bid-Ask Spreads in the Foreign Exchange Markets', mimeo, Brussels.

Bruno, M., and Sachs, J. (1985) *Economics of Worldwide Stagflation*, Oxford: Basil Blackwell.

Buchanan, J., and Tullock, T. (1962) *The Calculus of Consent*, Ann Arbor: University of Michigan Press.

Buiter, W., and Kletzer, K. (1990) 'Reflections on the Fiscal Implications of a Common Currency', CEPR Discussion Paper, no. 418.

—— —— (1991) 'Reflections on the Fiscal Implications of a Common Currency', in A. Giovannini and C. Mayer (eds.), *European Financial Integration*, Cambridge: CUP.

—— Corsetti, G., and Roubini, N. (1993) 'Sense and Nonsense in the Treaty of Maastricht', *Economic Policy*, 16.

Bundesministerium der Finanzen (1996) *Finanzbericht 1997*, Bonn.

Cagan, P. (1956) 'The Monetary Dynamics of Hyperinflation', in M. Friedman (ed.), *Studies in the Quantity Theory of Money*, Chicago: Chicago University Press.

Calmfors, L., and Driffill, J. (1988) 'Bargaining Structure, Corporatism and Macroeconomic Performance', *Economic Policy*, 6: 13–61.

Canzoneri, M., Valles, J., and Vinals, J. (1996) 'Do Exchange Rates Have to Address International Macroeconomic Imbalances?', CEPR Discussion Paper, no. 1498.

Carlin, W., and Soskice, D. (1990) *Macroeconomics and the Wage Bargain*, Oxford: OUP.

Cohen, D. (1983) 'La Coopération monétaire européenne: du SME à l'Union monétaire', *Revue d'économie financière*, 8/9.

—— and Wyplosz, C. (1989) 'The European Monetary Union: An Agnostic Evaluation', unpub. typescript.

Collins, S. (1988) 'Inflation and the European Monetary System', in Giavazzi *et al.* (1988) *The European Monetary System*, Cambridge: CUP.

Committee on the Study of Economic and Monetary Union (the Delors Committee) (1989) *Report on Economic and Monetary Union in the European Community* (*Delors Report*) (with Collection of Papers), Luxemburg: Office for Official Publications of the European Communities.

Connolly, B. (1995) *The Rotten Heart of Europe: The Dirty War for Europe's Money*, London: Faber and Faber.

Corden, M. (1972) 'Monetary Integration', *Essays in International Finance*, no. 93, Princeton.

Costa, C. (1996) 'Exchange Rate Pass-Through: The Case of Portuguese Imports and Exports', unpub. typescript, University of Leuven.

Cukierman, A. (1992) *Central Bank Strategy, Credibility and Independence, Theory and Evidence*, Cambridge, Mass.: MIT Press.

David, P. (1985) 'CLIO and the Economics of QWERTY', *American Economic Review*, 75: 332–7.

Davis, S., Haltiwanger, J., and Schuh, S. (1996) *Job Creation and Destruction*, Cambridge, Mass.: MIT Press.

De Cecco, M., and Giovannini, A. (eds.) (1989) *A European Central Bank? Perspectives on Monetary Unification after Ten Years of the EMS*, Cambridge: CUP.

De Grauwe, P. (1975) 'Conditions for Monetary Integration: A Geometric Interpretation', *Weltwirtschaftliches Archiv*, 111: 634–46.

—— (1983) *Macroeconomic Theory for the Open Economy*, Aldershot: Gower.

—— (1987) 'International Trade and Economic Growth in the EMS', *European Economic Review*, 31: 389–98.

—— (1990) 'The Cost of Disinflation and the European Monetary System', *Open Economics Review*, 1: 147–73.

—— (1991) 'Is the EMS a DM-Zone?', in A. Steinherr and D. Weiserbs (eds.), *Evolution of the International and Regional Monetary Systems*, London: Macmillan.

—— and Heens, H. (1993) 'Real Exchange Rate Variability in Monetary Unions', *Recherches Économiques de Louvain*, 59/1–2.

—— and Peeters, T. (eds.) (1989) *The ECU and European Monetary Integration*, London: Macmillan.

—— and Vanhaverbeke, W. (1990) 'Exchange Rate Experiences of Small EMS Countries. The Cases of Belgium, Denmark and the Netherlands', in V. Argy and P. De Grauwe (eds.), *Choosing an Exchange Rate Regime*, Washington, DC: International Monetary Fund.

—— —— (1993) 'Is Europe an Optimum Currency Area?: Evidence from Regional Data', in P. R. Masson and M. P. Taylor (eds.), *Policy Issues in the Operation of Currency Unions*, Cambridge: CUP.

de Haan, J., and Eijffinger, S. (1994) 'De Politieke Economie van Centrale Bank Onafhankelijkheid', *Rotterdamse Monetaire Studies*, no. 2.

Demopoulos, G., Katsimbris, G., and Miller, S. (1987) 'Monetary Policy and Central Bank Financing of Government Budget Deficits: A Cross-Country Comparison', *European Economic Review*, 31: 1023–50.

Dominguez, K., and Frankel, J. (1993) *Does Foreign Exchange Intervention Work?*, Washington, DC: Institute for International Economics.

Dornbusch, R. (1976) 'Money and Finance in European Integration', mimeo, Cambridge, Mass.: MIT.

—— (1988) 'The European Monetary System, the Dollar and the Yen', in Giavazzi *et al.* (1988) *The European Monetary System*, Cambridge: CUP.

—— and Fischer, S. (1978) *Macroeconomics*, New York: McGraw-Hill.

Driffill, J. (1988) 'The Stability and Sustainability of the European Monetary System with Perfect Capital Markets', in Giavazzi *et al.* (1988) *The European Monetary System* Cambridge: CUP.

EC Commission (1977) Report of the Study Group on the Role of Public Finance in European Integration (MacDougall Report), Brussels.

—— (1988) 'The Economics of 1992', *European Economy*, 35.

—— (1990) 'One Market, One Money', *European Economy*, 44.

—— (1993) 'Stable Money—Sound Finances, Community Public Finance in the Perspective of EMU', *European Economy*, 53.

Eichengreen, B. (1990) 'Is Europe an Optimum Currency Area?', CEPR Discussion Paper, no. 478.

—— and von Hagen, J. (1995) 'Fiscal Policy and Monetary Union: Federalism, Fiscal Restrictions and the No-Bailout Rule', CEPR Discussion Paper, no. 1247.

—— and Wyplosz, C. (1993) 'The Unstable EMS', CEPR Discussion Paper, no. 817.

Eijffinger, S. and Schaling, E. (1995) 'The Ultimate Determinants of Central Bank Independence', in S. Eijffinger and H. Huizinga (eds.), *Positive Political Economy: Theory and Evidence*, New York: Wiley and Sons.

Engel, C., and Rogers, J. (1995) 'How Wide is the Border?', International Finance Discussion Paper, no. 498, Washington, DC: Board of Governors of the Federal Reserve System.

Erkel-Rousse, H., and Melitz, J. (1995) 'New Empirical Evidence on the Costs of Monetary Union', CEPR Discussion Paper, no. 1169.

European Commission (various years) *Car Prices within the European Union*, Brussels.

Fischer, S. (1982) 'Seigniorage and the Case for a National Money', *Journal of Political Economy*, 90: 295–307.

Frankel, J. and Rose, A. (1996) 'The Endogeneity of the Optimum Currency Area Criteria', NBER Discussion paper, no. 5700.

Fratianni, M. (1988) 'The European Monetary System: How Well has it Worked? Return to an Adjustable-Peg Arrangement', *Cato Journal*, 8: 477–501.

—— and Peeters, T. (eds.) (1978) *One Money for Europe*, London: Macmillan.

—— and von Hagen, J. (1990) 'German Dominance in the EMS: The Empirical Evidence', *Open Economies Review*, 1: 86–7.

—— —— (1992) *The European Monetary System and European Monetary Union*, Boulder, Colo.: Westview Press.

—— —— and Waller, C. (1992) 'The Maastricht Way to EMU', *Essays in International Finance*, no. 187, Princeton.

Friedman, M. (1967) 'The Role of Monetary Policy', *American Economic Review* 58: 1–17.

—— (1969 edn.) 'The Optimum Quantity of Money', in M. Friedman (ed.), *The Optimum Quantity of Money and Other Essays*, Chicago: Aldine.

Froot, R., and Obstfeld, M. (1989) 'Exchange Rate Dynamics under Stochastic Regime Shifts', NBER Working Paper, no. 2835.

Giavazzi, F. and Giovannini, A. (1989) *Limiting Exchange Rate Flexibility: The European Monetary System*, Cambridge, Mass.: MIT Press.

—— and Pagano, M. (1985) 'Capital Controls and the European Monetary System', in *Capital Controls and Foreign Exchange Legislation*, Euromobiliare, Occasional Paper 1.

—— —— (1988) 'The Advantage of Tying One's Hands: EMS Discipline and Central Bank Credibility', *European Economic Review*, 32: 1055–82.

—— —— (1990) 'Can Severe Fiscal Contractions be Expansionary? Tales of Two Small European Countries', *NBER Macroeconomic Annual*: 75–111.

—— and Spaventa, L. (1990) 'The "New" EMS', in P. De Grauwe and L. Papademos (eds.), *The European Monetary System in the 1990s*, London: Longman.

—— Micossi, S., and Miller, M. (eds.) (1988) *The European Monetary System*, Cambridge: CUP.

Giersch, H. (1973) 'On the Desirable Degree of Flexibility of Exchange Rates', *Weltwirtschaftliches Archiv*, 109: 191–213.

Giovannini, A., and Spaventa, L. (1991) 'Fiscal Rules in the European Monetary Union: A No-Entry Clause', CEPR Discussion Paper, no. 516.

Goodhart, C. (1989) '*The Delors Report*: Was Lawson's Reaction Justifiable?', London: Financial Markets Group, London School of Economics (unpub.).

Gordon, R. (1996) 'Problems in the Measurement and Performance of Service-Sector Productivity in the US', NBER Working paper, no. 5519.

Grilli, V. (1989) 'Seigniorage in Europe', in De Cecco and Giovannini (1989).

—— Masciandro, D., and Tabellini, G. (1991) 'Political and Monetary Institutions and Public Financial Policies in the Industrial Countries', *Economic Policy*, 13: 341–92.

Gros, D. (1990) 'Seigniorage and EMS Discipline', in P. De Grauwe and L. Papademos (eds.), *The European Monetary System in the 1990s*, London: Longman.

—— (1995) 'Towards a Credible Excessive Deficits Procedure', Brussels: Centre for European Policy Studies.

—— (1996) 'A Reconsideration of the Optimum Currency Approach: The Role of External Shocks and Labour Mobility', Brussels: Centre for European Policy Studies.

—— and Lannoo, K. (1996) *The Passage to the Euro*, CEPS Working Party Report, no. 16.

—— and Thygesen, N. (1988) 'The EMS: Achievements, Current Issues and Directions for the Future', CEPS Paper, no. 35, Brussels: Centre for European Policy Studies.

—— —— (1992) *European Monetary Integration: From the European Monetary System towards Monetary Union*, London: Longman.

Grubb, D., Jackman, R., and Layard, R. (1983) 'Wage Rigidity and Unemployment in OECD Countries', *European Economic Review*.

Hammond, G., and von Hagen, J. (1993) 'Regional Insurance against Asymmetric Shocks: An Empirical Study for the European Community', University of Mannheim, typescript.

Hayek, F. (1978) *Denationalization of Money*, London: Institute of Economic Affairs.

HM Treasury (1989) *An Evolutionary Approach to Economic and Monetary Union*, London: HMSO.

—— (1990) 'The Hard ECU Proposal', mimeo.

Holtfrerich, C.-L. (1989) 'The Monetary Unification Process in Nineteenth-Century Germany: Relevance and Lessons for Europe Today', in De Cecco and Giovannini (1989).

Ingram, J. (1959) 'State and Regional Payments Mechanisms', *Quarterly Journal of Economics*, 73: 619–32.

International Monetary Fund (1984) 'Exchange Rate Volatility and World Trade: A Study by the Research Department of the International Monetary Fund', *Occasional Papers*, no. 28.

Italianer, A., and Vanheukelen, M. (1992) 'Proposals for Community Stabilization Mechanisms' in 'The Economics of Community Public Finances', *European Economy*, Special Issue.

Jozzo, A. (1989) 'The Use of the ECU as an Invoicing Currency', in P. De Grauwe and T. Peeters (eds.), *The ECU and European Monetary Integration*, London: Macmillan.

Kaldor, N. (1966) *The Causes of the Slow Growth of the United Kingdom*, Cambridge: CUP.

Kenen, P. (1969) 'The Theory of Optimum Currency Areas: An Eclectic View', in R. Mundell and A. Swobodaa (eds.), *Monetary Problems of the International Economy*, Chicago: University of Chicago Press.

Klein, B. (1978) 'Competing Monies, European Monetary Union and the Dollar', in Fratianni and Peeters (1978).

Krugman, P. (1987) 'Trigger Strategies and Price Dynamics in Equity and Foreign Exchange Markets', NBER Working Paper, no. 2459.

—— (1989) 'Differences in Income Elasticities and Trends in Real Exchange Rates', *European Economic Review*, 33: 1031–47.

—— (1990) 'Policy Problems of a Monetary Union', in P. De Grauwe and L. Papademos (eds.), *The European Monetary System in the 1990s*, London: Longman.

—— (1991) *Geography and Trade*, Cambridge, Mass.: MIT Press.

—— (1993) 'Lessons of Massachusetts for EMU', in F. Torres and F. Giavazzi (eds.), *Adjustment and Growth in the European Monetary Union*, London: CEPR, and Cambridge: CUP.

Kydland, F., and Prescott, E. (1977) 'Rules Rather than Discretion: The Inconsistency of Optimal Plans', *Journal of Political Economy*, 85.

Lamfalussy, A. (1989) 'Macro-coordination of Fiscal Policies in an Economic and Monetary Union', in Committee on the Study of Economic and Monetary Union (1989).

Leiderman, L., and Svensson, L. (eds.) (1995) *Inflation Targets*, London: CEPR.

Lomax, D. (1989) 'The ECU as an Investment Currency,' in P. De Grauwe and T. Peeters (eds.), *The ECU and European Monetary Integration*, London: Macmillan.

Ludlow, P. (1982) *The Making of the European Monetary System*, London: Butterworths.

Maastricht Treaty (Treaty on European Union) (1992) CONF-UP-UEM 2002/92, Brussels, 1 Feb.

Mastropasqua, C., Micossi, S., and Rinalid, R. (1988) 'Interventions, Sterilization and Monetary Policy in the EMS Countries (1979–1987)', in Giavazzi *et al.* (1988).

McDonald, I., and Solow, R. (1981) 'Wage Bargaining and Employment', *American Economic Review*, 71: 896–908.

McKinnon, R. (1963) 'Optimum Currency Areas', *American Economic Review*, 53: 717–25.

—— (1996) *Default Risk in Monetary Unions*, background report for the Swedish Government Commission on EMU, Stockholm.

Mélitz, J. (1985) 'The Welfare Cost of the European Monetary System', *Journal of International Money and Finance*, 4: 485–506.

—— (1988) 'Monetary Discipline, Germany and the European Monetary System: A Synthesis', in Giavazzi *et al.* (1988).

—— and Vori, S. (1993) 'National Insurance against Unevenly Distributed Shocks in a European Monetary Union', *Recherches Économiques de Louvain*, 59: 1–2.

Micossi, S., and Tullio, G. (1991) 'Fiscal Imbalances, Economic Distortions, and the Long Run Performance of the Italian Economy', OCSM Working Paper, no. 9, LUISS.

Moesen, W., and Van Rompuy, P. (1990) 'The Growth of Government Size and Fiscal Decentralization', Paper prepared for the IIPF Congress, Brussels.

Morales, A., and Padilla, A. J. (1994) 'Designing Institutions for International Monetary Co-operation', unpub., Madrid: CEMFI.

Mundell, R. (1961) 'A Theory of Optimal Currency Areas', *American Economic Review*, 51:

Myrdal, G. (1957) *Economic Theory and Underdeveloped Regions*, New York: Duckworth.

Neumann, M. (1990) 'Central Bank Independence as a Prerequisite of Price Stability', mimeo, University of Bonn.

—— and von Hagen, J. (1991) 'Real Exchange Rates within and between Currency Areas: How far away is EMU?', Discussion Paper, Indiana University.

Obstfeld, M. (1986) 'Rational and Self-Fulfilling Balance of Payments Crises', *American Economic Review*, 76: 72–81.

OECD (1990) *Economic Outlook*, Paris.

Padoa-Schioppa, T. (1988) 'The European Monetary System: A Long-Term View', in Giavazzi *et al.* (1988).

Parkin, M., and Bade, R. (1988) *Modern Macroeconomics*, 2nd edn., Oxford: Philip Allan.

Persson, T., and Svensson, L. (1989) 'Why a Stubborn Conservative would Run a Deficit: Policy with Time Inconsistent Preferences', *Quarterly Journal of Economics*, 104.

—— and Tabellini, G. (1996) Monetary Cohabitation in Europe', NBER Working Paper, no. 5532.

Phelps, E. (1968) 'Money-Wage Dynamics and Labour Market Equilibrium', *Journal of Political Economy*, 76: 678–711.

Poole, W. (1970) 'Optimal Choice of Monetary Policy Instruments in a Simple Stochastic Macro Model', *Quarterly Journal of Economics*, 85.

Portes, R. (1989) 'Macroeconomic Policy Coordination and the European Monetary System', CEPR Discussion Paper, no. 342.

—— (1993) 'EMS and EMU After the Fall', *World Economy*, 16: 1–16.

Posen, A. (1994) 'Is Central Bank Independence the Result of Effective Opposition to Inflation? Evidence of Endogenous Monetary Policy Institutions', mimeo, Harvard University.

Rogoff, K. (1985*a*) 'Can Exchange Rate Predictability be Achieved without Monetary Convergence? Evidence from the EMS', *European Economic Review*, 28: 93–115.

—— (1985*b*) 'The Optimal Degree of Commitment to an Intermediate Monetary Target', *Quarterly Journal of Economics*, 100: 1169–90.

Roll, E. (1993) 'Independent and Accountable: A New Mandate for the Bank of England', Report of an Independent Panel chaired by Eric Roll, London: CEPR.

Romer, P. (1986) 'Increasing Returns and Long-Term Growth', *Journal of Political Economy*.

Rose, A., and Svensson, L. (1993) 'European Exchange Rate Credibility Before the Fall', NBER Working Paper, no. 4495.

Roubini, N., and Sachs, J. (1989) 'Government Spending and Budget Deficits in the Industrial Countries', *Economic Policy*, 11: 100–32.

Russo, M., and Tullio, C. (1988) 'Monetary Policy Coordination within the European Monetary System: Is there a Rule?' in Giavazzi *et al.* (1988).

Sachs, J., and Sala-i-Martin, X. (1989) 'Federal Fiscal Policy and Optimum Currency Areas', Harvard University Working Paper, Cambridge, Mass.

—— and Wyplosz, C. (1986) 'The Economic Consequences of President Mitterand', *Economic Policy*, 2.

Sachverständigenrat zur Begutachtung der Gesamtwirtschaftlichen Entwicklung, 'Sondergutachten', 20 Jan. 1990.

Salin, P. (ed.) (1984) *Currency Competition and Monetary Union*, The Hague: Martinus Nijhoff.

Sannucci, V. (1989) 'The Establishment of a Central Bank: Italy in the Nineteenth Century', in De Cecco and Giovannini (1989).

Sargent, T., and Wallace, N. (1981) 'Some Unpleasant Monetarist Arithmetic', *Federal Reserve Bank of Minneapolis Quarterly Review*, 5: 1–17.

Shapiro, M., and Wilcox, D. (1996) 'Mismeasurement in the Consumer Price Index: An Evaluation', NBER Working Paper, no. 5590.

Steinherr, A. (1989) *Concrete Steps for Developing the ECU*, Report prepared by the ECU Banking Association Macrofinancial Study Group, Brussels.

Stiglitz, J., and Weiss, A. (1981) 'Credit Rationing in Markets with Imperfect Information', *American Economic Review*, 71: 393–410.

Straubhaar, T. (1988) 'International Labour Migration within a Common Market: Some Aspects of the EC Experience', *Journal of Common Market Studies*, 27.

Svensson, L. (1992) 'The Foreign Exchange Risk Premium in a Target Zone with a Devaluation Risk', *Journal of International Economics*, 33: 21–40.

—— (1995) 'Optimal Inflation Targets, Conservative Central Banks, and Linear Inflation Contracts', CEPR Discussion Paper, no. 1249.

Tavlas, G. (1993) 'The "New" Theory of Optimum Currency Areas', *World Economy*, 33: 663–82.

Thygesen, N. (1988) 'Decentralization and Accountability within the Central Bank: Any Lessons from the US Experience for the Potential Organization of a European Central Banking Institution?' (with comment by Jean-Jacques Rey), in P. De Grauwe and T. Peeters (eds.), *The ECU and European Monetary Integration*, Basingstoke: Macmillan.

—— (1989) 'A European Central Banking System: Some Analytical and Operational Considerations', in Collection of Papers of Delors Report, 157–75.

Tower, E., and Willett, T. (1976) 'The Theory of Optimum Currency Areas and Exchange Rate Flexibility', *Special Papers in International Finance*, no. 11, Princeton, NJ: Princeton University.

Turnovsky, S. (1984) 'Exchange Market Intervention under Alternative Forms of Exogenous Disturbances', *Journal of International Economics*, 17: 279–97.

Ungerer, H., Evans, O., and Byberg, P. (1986) 'The European Monetary System: Recent Developments', International Monetary Fund Occasional Paper, no. 48.

van der Ploeg, F. (1991) 'Macroeconomic Policy Coordination during the Various Phases of Economic and Monetary Integration in Europe', in EC Commission, *European Economy*, Special Edn., 1.

Van Neder, N., and Vanhaverbeke, W. (1990) 'The Causes of Price Differences in the European Car Markets', International Economics Discussion Paper, University of Leuven.

Van Rompuy, P., Abraham, F., and Heremans, D. (1991) 'Economic Federalism and the EMU', in EC Commission, *European Economy*, Special Edn., 1.

van Ypersele, J. (1985) *The European Monetary System*, Cambridge: Woodhead.

Vaubel, R. (1978) *Strategies for Currency Unification*, Kieler Studien, no. 156, Tübingen: Mohr & Siebeck.

—— (1989) 'Uberholte Glaubenssätze', *Wirtschaftsdienst*, 69: 276–9.

—— (1990) 'Currency Competition and European Monetary Integration', *Economic Journal*, 100: 936–46.

von Hagen, J. (1991) 'A Note on the Empirical Effectiveness of Formal Fiscal Restraints', *Journal of Public Economics*, 44: 199–210.

—— (1991) 'Fiscal Arrangements in a Monetary Union. Evidence from the US', Working Paper, Indiana University.

—— and Fratianni, M. (1990) 'Asymmetries and Realignments in the EMS', in P. De Grauwe and L. Papademos (eds.), *The European Monetary System in the 1990s*, London: Longman.

—— and Hammond, G. (1995) 'Regional Insurance against Asymmetric Shocks: An Empirical Study for the European Community', CEPR Discussion Paper, no. 1170.

—— and Lutz, S. (1996) 'Fiscal and Monetary Policy on the Way to EMU', *Open Economies Review*, 7: 299–325.

Walsh, C. (1995) 'Optimal Contracts for Independent Central Banks', *American Economic Review*, 85: 150–67.

Walters, A. (1986) *Britain's Economic Renaissance*, Oxford: OUP.

Weber, A. (1990) 'European Economic and Monetary Union and Asymmetric and Adjustment Problems in the European Monetary System: Some Empirical Evidence', University of Siegen Discussion Paper, no. 9–90.

Wickens, M. (1993) 'The Sustainability of Fiscal Policy and the Maastricht Conditions', London Business School Discussion Paper, no. 10–93.

Williamson, J. (1983) *The Exchange Rate System*, Washington, DC: Institute of International Economics.

Winkler, B. (1994) 'Reputation for EMU: An Economic Defence of the Maastricht Criteria', unpub. ms., European University Institute.

Wyplosz, C. (1989) 'Asymmetry in the EMS: Intentional or Systemic?', *European Economic Review*, 33: 310–20.

—— (1991) 'Monetary Union and Fiscal Policy Discipline', in EC Commission, *European Economy*, Special Edn., 1.

Zimmerman, H. (1989) 'Fiscal Equalization between States in West Germany', *Government and Policy*, 7: 385–93.

INDEX